Cyberterrorism

The Legal and Enforcement Issues

Cyberterrorism

The Legal and Enforcement Issues

Pardis Moslemzadeh Tehrani
University of Malaya, Malaysia

World Scientific

NEW JERSEY • LONDON • SINGAPORE • BEIJING • SHANGHAI • HONG KONG • TAIPEI • CHENNAI • TOKYO

Published by

World Scientific Publishing Europe Ltd.

57 Shelton Street, Covent Garden, London WC2H 9HE

Head office: 5 Toh Tuck Link, Singapore 596224

USA office: 27 Warren Street, Suite 401-402, Hackensack, NJ 07601

Library of Congress Cataloging-in-Publication Data
Names: Tehrani, Pardis Moslemzadeh, author.
Title: Cyberterrorism : the legal and enforcement issues / by Pardis Moslemzadeh Tehrani (University of Malaya, Malaysia).
Description: New Jersey : World Scientific, 2017. | Includes bibliographical references.
Identifiers: LCCN 2016039137 | ISBN 9781786342126 (hc : alk. paper)
Subjects: LCSH: Cyberterrorism--Law and legislation. | Information warfare (International law) | Internet and terrorism.
Classification: LCC KZ7225.C93 T44 2017 | DDC 345/.0268--dc23
LC record available at https://lccn.loc.gov/2016039137

British Library Cataloguing-in-Publication Data
A catalogue record for this book is available from the British Library.

Desk Editors: Dr. Sree Meenakshi Sajani/Mary Simpson

Typeset by Stallion Press
Email: enquiries@stallionpress.com

PREFACE

There are not many publications on cyberterrorism due to it being a relatively new phenomena although cyberattacks have been on an uptrend around the globe recently. Such aggressive digital activities rekindle fears that cyberterrorism is an inevitable threat to every nation. Cyberterrorist attacks pose many problems to countries as they cause physical and virtual damage to their national infrastructures. Despite this, the basic concept of cyberterrorism remains unclear and countries still do not have adequate legislation to combat it although many have taken steps in that direction. They have introduced statutes, cyber security measures, and joined international organisations in response to this problem. Over the past two decades voluminous literature has been produced on the issues of cyberterrorism. Technology obliges us in a way to be connected by communications and information technologies are more critical than ever before. This book integrates technology and law taking into account the situation in the US, the UK, and Malaysia, as well as the contributions of international organisations and related legal statutes. It assists the reader in having a better understanding of the notion of cyberterrorism and statutes created to address it in three jurisdictions particularly in the US which has led the way in enacting cyberterrorism legislation. Cyberterrorism is a global problem and it needs global harmonisation to combat it involving all nations and public and private stakeholders. Such a massive and coordinated attack against critical national and international infrastructures requires a global response and although regional and bilateral agreements can be effective in some cases they are not adequate.

ABOUT THE AUTHOR

Pardis Moslemzadeh Tehrani is a Senior Lecturer in the Faculty of Law, University of Malaya. Pardis Moslemzadeh Tehrani completed her Ph.D. at National University of Malaya. Her research interests lie in the areas of Cyberterrorism, Human Rights, International Humanitarian Law, International Trade Law, Intellectual Property Law, Cloud Computing in Law, ASEAN Studies. Pardis has served on many conference and workshop program committees. She is involved in quite a number of journals on the Editorial Review Board such as *International Journal of Digital Crime and Forensics*, *Asian Journal of Humanities and Social Studies*, and *International Science Index Journal*. She has served as an International Scientific Member of World Academy of Science, Engineering and Technology, and Asia Pacific Association of Technology and Society. Pardis has widely published in peer-reviewed journals and presented papers in a number of national and international conferences.

CONTENTS

ABSTRACT

Advancing technology has spawned new types of crime. Traditional crimes have migrated to the internet. Cyberterrorism is a new kind of crime distinct from cyber crime and has emerged as one of the most complex phenomena of this century that poses a significant threat to nations. Various international organisations have been established and a more comprehensive consensus among countries and organisations is needed to address the issue of cyberterrorism. This book investigates cyberterrorism from the aspect of international and domestic legal mechanisms. It seeks to establish a comprehensive definition of cyberterrorism and examines the various anti-terrorism legislations and measures taken by nations to protect themselves including measures taken to prosecute cyber criminals. Employing qualitative methodology, this study uses both analytical and case studies to compare illegal and legal usage of cyber space in the discussed countries. Data were analysed via analytic methods to discover casual factors that affects outcome. The findings show that there is an urgent need for a global consensus to counter cyberterrorism by harmonising their various legislations. States must cooperate to share information in order to apprehend attackers and sentence them. While individual countries are keen to implement their own laws to handle the problem, a better approach would be to build a framework for consensus among the international community. Owing to the trans-border nature of the internet and the reality of cyberterrorism, universal jurisdiction is the most suitable option to deter cyberterrorism. The existence of mutual legal assistance mechanisms, extradition treaties as well as bilateral and multilateral arrangements among countries will allow for a more effective response especially in prosecuting cyber terrorists. The existing international conventions and other legislations regarding the harmonisation of national laws and international cooperation are applicable to the misuse of the internet for terrorist purposes. The only issue is the existence of terrorist-specific gaps in computer-specific conventions and computer-specific gaps in terrorist-specific conventions.

LIST OF CASES MENTIONED

Barker v. R case
Bensusan v. King
Berger and Katz
Brahim Benmerzouga and Baghdad Meziane v. McLaughlin
Christopher Pile case
Code Red i and Code Red ii
Doe v. Holder
Director of Public Prosecutions v. Bignell
Director of Public Prosecutions v. David Lennon
Director of Public Prosecutions v. Murdoch case
Erie Railroad v. Tompkins
Estonia case
Factor v. Laubenheimer
Helicopteros Nacionales de Colombia v. Hall
I Love You Virus case
International Shoe Case
Invita case
Love Bug Virus case
McKinnon case

LIST OF STATUTES

Computer Misuse Act of United Kingdom 1990
Computer Crime Act of Malaysia 1997
Criminal Damage Act of Malaysia 1971
Criminal Procedure Code of Malaysia 1973
Cyber Security Enhancement Act of the United States 2002
Electronic Communications Privacy Act (ECPA) of the United States 1986
Emergency Act of the United States 2009
Federal Wiretap Act of the United States 1968
Foreign Intelligence Surveillance Act of 1978 (FISA) United States
Internal Security Act of Malaysia 1960
Omnibus Crime Control and Safe Streets Act of the United States 1968
Patriot Act of the United States 2001
Penal Code of Malaysia
Police and Justice Act of the United Kingdom 2006
Regulation of Investigatory Powers Act of the United Kingdom 2000
Terrorism Act of the United Kingdom 2000
The Anti-Terrorism, Crime and Security Act of the United Kingdom 2001
The Computer Fraud and Abuse Act of the United States (CFAA)
The Security Offences Act of Malaysia 2012
United States Code

LIST OF INTERNATIONAL INSTRUMENTS

The Council of Europe Convention of Cybercrime (2001)
Convention on the Prevention of Terrorism
United Nations Security Council Resolution 1373 (2001)
United Nations Charter
North Atlantic Treaty Organization (NATO)
United Nations Security Council Resolution 1373
International Telecommunications Union (ITU)
The United Nations Office on Drug and Crime (UNODC)
The Organization for Security and Cooperation in Europe
Interpol
The Council of Europe, Asia-Pacific Economic Cooperation (APEC)
North Atlantic Treaty Organization
International Multilateral Partnership against Cyber Terrorism (IMPACT)
The Organization for Economic Cooperation and Development

CHAPTER I

DEFINITIONAL ISSUES RELATING TO CYBERTERRORISM

1.1 INTRODUCTION

Today, we see that traditional crime has migrated to the internet. The types of crimes that people have been committing over the years, such as fraud and extortion and crimes of a similar nature have just moved on to the internet. Therefore, where people might have taken photos of you in a compromising position before and extorted you, now they will steal your information and extort you by threatening to release that information unless they are paid, or by destroying that information unless they are paid.

Cyberterrorism is a tempting proposition for a terrorist group, as they would require less people and fewer resources. Moreover, it enables the terrorist to remain unknown, since it is carried out far away from the actual location of the terrorist. Cyber terrorists can set up anywhere and remain anonymous. It consists of physical terrorism and cyberterrorism. Cyber terrorists exploit technology via the internet to implement their terrorist purposes. The idea of a "Digital World Trade Center Attack", possibly killing thousands and causing billions of dollars in damage, is very real.[1] It is a development of

[1] L. Carlos *et al.*, 'Cyber terrorism — A rising threat in the western hemisphere', *Albany Law Journal of Science and Technology* (2008) 18, p. 298.

terrorist capabilities provided by new technologies and networked organisations, which allow terrorists to conduct their operations with little or no physical risks to themselves.

Cyberterrorism is a new and somewhat vague concept. However, there has been much debate over this term. The debates arise from the question of whether cyberterrorism is a separate phenomenon, or just a facet of information warfare practised by terrorists.[2] Scholars around the world cannot come to a consensus on the term "cyberterrorism". The previously mentioned experts provide specific definitions from different perspectives. This conflict can be attributed to the fact that the term "cyberterrorism" has no universally accepted definition. In fact, the definition of cyberterrorism is one of the main issues in addressing cyberterrorism threats. First, the problem should be identified when an incident happens. Second, the problem should be categorised to respond to and deal with such case appropriately. Thus, future incidents can be prevented.

If the term is not defined, it might lead to misused application of a cyberterrorism definition by considering another event as being prejudicial to national security or public safety of the country. In addition, any cyber crime may be interpreted as an act of cyberterrorism. If there is no law and definition for cyberterrorism, how can it be punished since, without law there is no sin? Therefore, a precise, accurate definition of cyberterrorism is the most significant point in addressing the issues of cyberterrorism. As we know, cyberterrorism and terrorism are breaches of law for which there is no need for any special measures. Thus, to prepare a proper legal response, the definition of terrorism should be considered.

Consequently, most countries do not have a particular definition for cyberterrorism. Thus, according to the norms of international treaty law, two approaches exist. First, conventions must expand to cover new circumstances, either implicitly or explicitly. Second,

---

[2] US Army Training and Doctrine Command, *Cyber Terrorism and Cyber Operations*, 3rd Edn, 2005, p. 17.

according to Part III of the Vienna Convention on the Law of Treaties, a new treaty can be concluded to adjust the legal changes to the legal reality.[3] Similarly, some parts of international law norms may be conveyed to national laws and utilised. A country which does not have cyberterrorism legislation may apply terrorist attack and cyber-crime legislation to a terrorist cyber attack. Therefore, that legislation and its definitions are also important.

1.1.1 Objective of the Chapter

No consensus among cyberterrorism scholars exists regarding the universal definition of cyberterrorism. Therefore, the present study aims to define cyberterrorism and its related terms. This definition can be used to identify offenses, investigate cases, and prosecute offenders. The objectives of this chapter are:

(i) To establish a proper definition of cyberterrorism this can be used as a basis to prosecute cyberterrorism offenses.
(ii) To delineate the differences between cyberterrorism offenses and other related cyber crime.

1.2 THE DEFINITIONS OF IMPORTANT TERMINOLOGIES

This chapter deals with definitions of technologies that are important to the book as a whole. It begins with the definitions of cyber and cyber space and is followed by definitions of terrorism and cyberterrorism. The discussion is mainly based on the dictionary, technical, and legal definitions which attempt to look at the technologies from different perspectives.

---

[3] A. Cohen, 'Cyber terrorism: Are we legally ready?' *Journal of International Business and Law* (2010) 9(1), p. 13.

1.2.1 Cyber and Cyber Space

"Cyberterrorism" is closely related to the act of terrorism. Therefore, the terms cyber, terrorism, and cyberterrorism are all important to be defined.

A. Dictionary definition

i. *Cyber*

Webopedia is an online dictionary and search engine for internet-related terms and technical support. This website defines cyber as "a prefix used in a growing number of terms to describe new things that are being made possible by the spread of computers".[4] The common elements of all these definitions are a set of interconnected information systems and the human users that interact with these systems. The human factor is the most significant part of the term "cyber" because anything cyber is created by humans.

ii. *Cyber space*

'Cyber space' is defined in Webopedia as "A metaphor for describing the non-physical terrain created by computer systems. Online systems, for example, create a cyber space within which people can communicate with one another (via e-mail), do research, or simply window shop. Like physical space, cyber space contains objects (files, mail messages, graphics, etc.), and different modes of transportation and delivery".[5] Even with the fantastic growth in cyber space and Information Technology (IT) experienced to date, "we have travelled only up the initial and shallow part of the exponential curve".[6]

The term cyber space was first coined in a science-fiction novel that was written by William Gibson. He described cyber space as a "consensual hallucination experienced daily by billions of legitimate operators, in every nation, by children being taught mathematical concepts. A graphic representation of data abstracted from banks of

[4] Available at: http://www.webopedia.com/TERM/C/cyber.html (28 Mar 2013).

[5] Available at: http://www.webopedia.com/TERM/c/cyberspace.html (2 Oct 2010).

[6] T. Altobelli, 'From the world congress: Cyber-abuse-a new worldwide threat to children's rights', *Family Court Review* (2010) 48, pp. 462–463.

every computer in the human system. Unthinkable complexity".[7] There are special programs creating specific cyber spaces, which are mostly used in computer games.[8] In order to explore cyber space, it is not necessary to perform any physical movements; shaking a mouse or touching any key on a keyboard is enough.

B. Technical definition

i. *Cyber*

Technically the term 'cyber' is defined as "A prefix taken from the word cybernetics (Greek *kybernan*, to steer or govern), and attached to other words having to do with computers and communication".[9] It is a prefix used to describe a person, thing, or idea as part of the computer and information age.[10]

C. Legal perspective

i. *Cyber space*

Cyber space is relatively new, and attempts to apply the rule of ordinary space to this new space have failed to address its unique problems.[11] This section aims to define cyber space using legal approaches related to cyber attacks. Defining cyber space is a substantial issue in cyber legal conflicts, since once it is defined explicitly, legal cases that occur in cyber space can be settled in a better way, because, it can give decision makers the proper means to analyse every case in different fields.[12] Defining cyber space in a proper way may lead to

---

[7] W. Gibson, *Neuromancer*, Ace Books Publication, New York, 1984, pp. 50–55.

[8] Webopedia Dictionary. Available at: http://www.webopedia.com/Term/c/cyberspace.html (2 Oct 2010).

[9] A. Seaton, *Federal Chambers Advanced English Dictionary*, 1st Edn, Federal Publications, Singapore, 2000, p. 190.

[10] Available at: http://searchsoa.techtarget.com/definition/cyber (28 Mar 2013).

[11] Thomas C. Folsom, 'Defining cyber space (finding real virtue in the place of virtual reality)', *Tulane Journal of Technology and Intellectual Property* (2007) 75(9). Available at: http://lexisnexis.com (20 Feb 2012).

[12] T. C. Folsom, 'Defining cyber space', *Tulane Journal of Technology and Intellectual Property* (2007) 75(9), pp. 230–232. Available at: http://lexisnexis.com (20 Feb 2012).

the resolution of many legal problems.[13] Cyber space is different from real space. Some scholars argue that the uncritical adoption of cyber space produces political consequences, as if divergent approaches indicate that cyber space is separate from other spaces, however, it is connected to real space by the consequences that are produced.[14]

Furthermore, for cyber space to be defined, this definition must be viewed from the angle of attacks that are attributed to cyber space. In defining cyber space, many factors must be taken into account, since it is a new medium which is capable of receiving threats and attacks that differ from traditional ones. The first factor is the speed with which an attack can take place in cyber space. A second factor is the result of a cyber space attack that is similar to that of weapons of mass destruction. The other factors are the cost of and the party responsible for the attack. In contrast with conventional methods of threats, the cost of conducting a cyber attack is low.[15] As the terrain of cyber space differs from physical space, the threats therein fundamentally differ from threats in physical space since the attack merely needs to be implemented from one computer to another computer. The result of such an attack is not kinetic and mostly involves manipulation of data or disruption of service. Even if not constrained to physical borders, the result may be the same as that of weapons of mass destruction. In fact, generally, a cyber space attack is much more destructive than the traditional method.

The European Convention on Cybercrime is the only substantial international legal document that addresses cyber space. The approach of the Convention in addressing such attacks occurring in cyber space is via criminal law. Furthermore, according to Article 51 of the United

---

[13] Thomas C. Folsom, 'Defining cyber space (Finding real virtue in the place of virtual reality)', *Tulane Journal of Technology and Intellectual Property* (2007) 75(9). Available at: http://lexisnexis.com (20 Feb 2012).

[14] J. E. Cohen, 'Cyberspace as/and space', *The Columbia Law Review* (2007) 107(1), p. 220. Available at: http://lexisnexis.com (25 Aug 2011).

[15] Major Graham H. Todd, 'Armed attack in cyber space: Deterring asymmetric warfare with asymmetric definition', *The Air Force Law Review* (2009) 64(65). Available at: http://lexisnexis.com (20 Feb 2012).

Nations (UN) Charter, parties have the right to act in self-defence.[16] The domain of cyber space is evidently not as physically transparent as the domains of air and space. If one were to try and "fire" 1 million volts at a cyber space target, an excessive number of information nodes (such as routers and switches) along the internet pathway would become the unintended recipient of the "electromagnetic missile".

The 1 million volt "electromagnetic missile" would cause collateral damage to these nodes and would be worn down long before it reaches its intended target. Cyber space attacks, similar to criminal hacks, are designed to affect electromagnetic data in various ways, consequently affecting the adversary's cyber space to create an operational advantage for the attacker. Thus, the basic legal framework of the Cybercrime Convention is beneficial in the context of military operations in cyber space.[17] The Dictionary of Military and Associated Terms defines cyber space as "the notional environment in which digitized information is communicated over computer networks". Once cyber space is defined, legal problems in several fields can then be better handled by adding two new factors to the otherwise ordinary analysis in whichever field is implicated: nature and place of use.[18]

D. Perspective in international documentation

i. *Cyber space*

In solving the conflicts that occur in cyber space, international law seems to be the best medium to address issues relate to cyber space. However, applying international law to attacks from cyber space to resolve conflicts and international disputes is a most difficult job,

---

[16] Major Graham H. Todd, 'Armed attack in cyber space: Deterring asymmetric warfare with asymmetric definition'.

[17] M. H. Todd, 'Armed attack in cyber space: Deterring asymmetric warfare with asymmetric definition'.

[18] Thomas C. Folsom, 'Defining cyber space (finding real virtue in the place of virtual reality)', *Tulane Journal of Technology and Intellectual Property* (2007) 75(9). Available at: http://lexisnexis.com (20 Feb 2012).

since it is hard to attribute old terms and norms to new technologies such as cyber space.[19]

Graham H. Todd, Chief of Operations Law at Eighth Air Force, Barksdale Air Force Base, Louisiana, as an international scholar in the US, defines cyber space as "an evolving man-made domain for the organization and transfer of data using various wavelengths of the electromagnetic spectrum. The domain is a combination of private and public property governed by technical rule sets designed primarily to facilitate the flow of information".[20] The European Commission provides a vague definition for cyber space as "the virtual space in which the electronic data of worldwide PCs circulate".[21]

As a result, most of the definitions that are provided for cyber space do not mention anything about the human factor. Different viewpoints explain a variety of definitions. The common factor in all definitions is the technological part: hardware, software, and data. Although all of these information sources must be transferred using personal computers, no reference to the human factor is made. The human factor is important in premier definitions about cyber space offered by Wiener and Gibson. In fact, humans created cyber space for human purposes. Similarly, these definitions do not consider the time factor. Time is absent from most definitions offered for cyber space. The notion of cyber space can be described in an appropriate manner and its complexity can be understandable if the time factor is also considered.[22]

The time factor also plays an important role, since other factors in relation to cyber space definitions can change or remain unchanged by

---

[19] G. H. Todd, 'Armed attack in cyber space: Deterring asymmetric warfare with an asymmetric definition', *The Air Force Law Review* (2009) 64(1), p. 70. Available at: http://lexisnexis.com (25 Aug 2011).

[20] G. H. Todd, 'Armed attack in cyberspace: Deterring asymmetric warfare with an asymmetric definition', p. 70.

[21] Information Society Website. Available at: http://ec.europa.eu/archives/ISPO/infocentre/glossary/i_glossary.html (28 Mar 2013).

[22] R. Ottis, P. Lorents, Cyberspace: Definition and Implications, *Proceeding of the International Conference on Information Warfare, Cooperative Cyber Defence Centre of Excellence*, Tallinn, Estonia, 2010, pp. 268–269.

the progression of the time. Put simply, all elements of cyber space are transformed over time. In addition, the span of time in cyber space can be short minutes, or even seconds. Offensive or defensive actions can be deployed rapidly. Therefore, attacks can be implemented in a short amount of time compared to those in physical space. A self-replicating worm can infect a huge part of cyber space in less than 1 minute. Such a case happened in 2003, when the SQL Slammer worm infected approximately 90% of hosts connected to the internet in less than 10 minutes.[23]

An overview of the definitions indicates that a complete definition must consider all these factors together in order to be effective. In this definition, the complete factors are technology, hardware, software, data, the human factor, and time. The NATO Cooperative Cyber Defence Centre of Excellence proposes a definition that seems to encompass all these factors. It defines cyber space as: "a time-dependent set of interconnected information systems and the human users that interact with these systems".[24]

Given the understanding of the threats and issues in cyber space, adopting a new law has thus become urgent. Describing cyber space is much easier than defining it. Cyber space goes beyond the description of the internet. At a very basic level, the internet is a vast collection of interconnected computer networks that enables the intermingled transmission of text, graphic, and sound files. An overview of existing definitions indicates that no common definition exists for cyber space, and the existing ones suffer from vague key points. The most appropriate definition for cyber space seems to be that offered by one of the scholars in the North Atlantic Treaty Organization (NATO) Cooperative Cyber Defense Center of Excellence. This scholar defined cyber space as "a time-dependent set of interconnected information systems and the human users that interact with these systems". In this definition, the human and time factors are included.

---

[23] D. Moore *et al.*, 'Inside the slammer worm', *IEEE Security and Privacy* (2003) 1(4), pp. 33–50.

[24] R. Ottis *et al.*, 'Cyber Space: Definition and implication', Cooperative Cyber Defence Centre of Excellence (2010) ICIW. Available at: http://www.mendeley.com/research/cyberspace-definition-implications (1 Mar 2012).

In conclusion, new technology needs new law. Particularly, if the new technology produces transformed relationships or changed technology and the existing law cannot address this properly, even if it destroys the enjoyment of the new technology in some situations. For instance, when automobiles were first produced, the related law did not exist and the existing law failed to handle the harm caused by automobiles. Then, the new problems arising from new relationships that were introduced by the new technology, new laws were adopted to address these problems.

1.2.2 Terrorism and Cyberterrorism

A. Dictionary definition

i. *Terrorism*

Black's Law Dictionary defines 'terrorism' as "the use or threat of violence to intimidate or cause panic, especially as a means of affecting political conduct".[25] Scholars and academics have also attempted to define 'terrorism', giving rise to a variety of definitions. However, these definitions may be broken down into five constituent elements:

(1) Violence committed by whatever means;
(2) Targeting innocent civilians;
(3) Intentionally causing violence with malicious disregard for its consequences;
(4) In order to cause fear, or coerce, or intimidate an enemy;
(5) For the purpose of realising a political, military, racial, ideological or religious goal.[26]

Some of the elements of this terrorism definition are troublesome. The definition of 'violence' is one of the elements which is normally associated with terrorism and perpetrated without excuse with the

[25] A. Brayan, Garner, *Black's Law Dictionary*, 7th Edn, Garner Publications United States, 1996, p. 1484.
[26] S. Tiefebrun, 'A semiotic approach to a legal definition of terrorism', *ISLA Journal of International and Comparative Law* (2003) 9, pp. 361–362.

aim to gain publicity for a cause. It is like civil disobedience in an extreme form through which the perpetrator seeks to gain publicity against an unjust law. Normally, in the case of state-sponsored terrorists, the publicity aspect is not seen, but it has been seen in the case of individual terrorists. Black's Law Dictionary defines 'violence' as including "unjust or unwarranted use of force, usually accompanied by fury, vehemence, or outrage; physical force unlawfully exercised with the intent to harm".[27]

However, the fact is that 'violence' can cover a broad range and has many forms and degrees of severity. For example, decided cases have held that violence need not be limited only to physical harm or injury, but also include picketing in a labour dispute carried out with misleading signs, false statements, erroneous publicity, and veiled threats. Further, while it may be said that an act is 'violent' only if it causes damage to persons and property,[28] violence in any form can inspire terror in both primary and secondary victims of that violence.

Another element that is troublesome is the term 'innocent civilians'. There has been no agreement as to who is covered under this category. There is debate that if an innocent civilian is killed during wartime, is this enough to label it as a terrorist act? Due to the established definitions used by the courts to determine the mental state of an accused, the element of intent is less troublesome. The element of 'fear' is problematic; however, since it is not "a legal term but a psychological phenomenon that is manifested by various signs and symptoms such as trembling, shaking, sweaty palms, etc. Intimidation, which was established as a tort in England as early as 1964, is unlawful coercion that produces harm".[29]

There are multiple reasons for acts of terrorism; they can include achieving political, military, racial, ideological, or religious aims. Thus, the overriding purpose of the act must be a necessary component of

---

[27] A. Brayan, *Black's Law Dictionary*, p. 1564.

[28] M. C. Bassiouni, 'International terrorism; Multilateral conventions', *Transnational Publishers* (2001) 14, p. 48.

[29] S. Tiefebrun, 'A semiotic approach to a legal definition of terrorism', *ISLA Journal of International and Comparative Law* (2003) 9, p. 362.

the definition. In regards to this, political, racial, ideological, and religious aims are not difficult to include in a definition of terrorism; it is the achievement of a 'military aim' that is controversial. If such a 'military aim' is included in a definition of terrorism, then this means terrorism cannot be used during a war. But the reality is that combatants (individuals, groups, or states) have committed terrorist acts during wars of national liberation.

B. Technical definition

i. *Cyberterrorism*

In this age of IT, terrorists have acquired the expertise to produce the most deadly combination of weapons and technology, against which the public has to be properly safeguarded in due course of time.[30] However, although cyberterrorism is a relatively new area of investigation, the underlying fundamentals have existed for centuries. It is only, the methods used by cyber terrorists and the motivation behind them that have changed. The initial hackers merely advocated the free sharing of information and they never harmed the data they accessed. However, this original 'hacker's ethic' does not seem to exist much anymore. Instead, with the advent of new technology, cyber terrorists have used these new advancements to advance their religious or political agendas. Cyberterrorism has now become a serious threat to both national and international security.[31]

Cyber terrorists now avail themselves of the latest forms of intelligence. Although "intelligence" can be said to be a wide-ranging term that includes all methods of covert communications and methods to unearth such communications, this old concept of the term is being overturned as the methods and motivation that drive cyber terrorists' hacking and code breaking have significantly changed over

---

[30] P. Dalal, 'Cybercrime and cyber terrorism: Preventive defense for cyberspace violations', *Computer Crime Research Center* (2006). Available at: http://www.crime-research.org/articles/1873/ (2 Apr 2010).

[31] P. W. Brunst, 'Legal aspect of cyber terrorism', in Centre of Excellence Defence against Terrorism. *Responses to Cyber Terrorism*, IOS Press, Ankara, Turkey, 2008, p. 67.

time.[32] However, intelligence-tracking methods have become outdated and cyber terrorists use new methods such as Brute Force, which is important in decryption of codes. Often, the cyber terrorist works as part of a group connected via the internet to attempt to break even the strongest encryption. In addition, they use Netbots to bombard websites in denial of service (DoS) attacks and flood the network with information, as well as shutting down websites by initiating distributed denial of service attacks (DDoS). The attacker targets other computers known as "zombies" or "bots" by using a large number of malware to infect a website or send spam to particular email addresses. Terrorist group attacks against electrical grids and pipeline systems use advanced types of DoS or DDoS. In recent times, cyber terrorists have launched DoS and DDoS attacks against the US, South Korea, and Estonia.

C. Legal perspective

i. *Definition derived from domestic legislations*

a. The US perspective

Previously, the US defined and classified international terrorism as a "crime" and provided legal tools to counter it. Recently, it has moved from a reactive technique to a proactive technique, since it now perceives terrorism as "an act of war" instead of just a 'crime'. Because of this, the US now uses newly-restructured law enforcement and intelligence agencies such as the Federal Bureau of Investigation (FBI) and the Central Intelligence Agency (CIA) in its war against terrorism, and these agencies each have their own definitions of terrorism.

The US Code defines "terrorism" as "a premeditated, politically motivated violence perpetrated against non-combatant targets by subnational groups or clandestine agents".[33] The code specifies the act, motive, actor, and victim, but does not address the legality of the

---

[32] K. Gable, 'Cyber-apocalypse now: Securing the internet against cyber terrorism and using universal jurisdiction as a deterrent', *Journal of Transnational Law* (2010) 43. Available at: http://lexisnexis.com (9 Jul 2010).

[33] Article 2656f (d)(2) United States Code.

violence. Notably, this definition excludes many dramatic examples of political violence against civilians. This definition has five components (i.e. premeditated, politically motivated, violent, targets being non-combatants, and performed by sub-national or clandestine agents) that must all be present to constitute an act of terrorism. For instance, the attacks directed against the World Trade Center and the Pentagon on 11 September 2001 (September 11 attacks) included all the components particularly because they were premeditated and politically motivated.

The US State Department has defined terrorism as "violence against non-combatants for the purpose of influencing public opinion" (Council on Foreign Relations [CFR]).[34] The former Deputy Chief of the CIA Counterterrorist Center has identified four elements that are common to all acts of terrorism. These are:

(i) Premeditated acts which are not simply acts born of rage.
(ii) Political in nature and designed to impact political structure.
(iii) Targeted at civilians and civilian installations.
(iv) Conducted by ad hoc groups as opposed to national armies.[35]

It has been accepted that the war on terrorism is bound to result in cyber attacks against US assets launched by terrorist groups, nation states that provide support for terrorists and hackers who sympathise with the terrorists. [36]

Following 11 September, 2001, when the US failed to respond to the cyber attack, it focused its energies on national-level laws that rely heavily on deterrence and prosecution of online terrorist activities. Thus, in the week following the September 11 attacks, the Patriot Act was passed by Congress to provide the necessary tools to address the

---

[34] Council on Foreign Relations. (n.d.) Terrorism: An introduction. Available at: http://www.terrorismanswers.com/terrorism/introduction.html (20 Jan 2013).

[35] A. Garner, Henry Campbell Black, *Black's Law Dictionary*, 9th Edn, West Publishers, London, 2009, p. 387.

[36] M. Vatis, 'Cyber-attacks during the war on terrorism', *Institute for Security Technology Studies at Dartmouth College* (2001) 5, p. 5. Available at: http://www.ists.dartmouth.edu (10 Jun 2001).

cyberterrorism issue, and to address inadequacies in the US' national homeland security. The Patriot Act makes it illegal for an individual to carry out offensive cyber attacks resulting in physical injury to national citizens, damage to US facilities, or to threaten public health or safety. The US' approach in criminalising cyberterrorism determines the penalties for a cyber attacker according to the level of harm caused. It reserves the maximum penalty of life imprisonment for only those acts of cyberterrorism that cause or attempt to cause death.

The Patriot Act's purpose was "to enhance the capability of the United States to deter, prevent, and thwart domestic and international acts of terrorism against United States nationals and interests". It was enacted for the investigation of terrorists and to bring them to justice. Following the passage of the US Patriot Act, which enables law enforcement agencies to investigate and prosecute potential threats to national security, the potential for hackers to be labelled as cyber terrorists and to face up to 20 years in prison has emerged. This law also gives broad powers to the government to track wireless phone calls, listen to voicemail, intercept email messages, and monitor computer use. President George W. Bush also created a new department in 2002 to gather and focus the government's efforts in facing the challenges of cyberterrorism.[37]

In new amendments to the USA Patriot Act, the term "domestic terrorism" has been redefined. Consequently, it includes mass destruction as well as assassination or kidnapping as a terrorist activity. The USA Patriot Act includes the new term "domestic terrorist" that could transform protesters into terrorists if they are associated with conduct that endangers human life.[38] The term "domestic terrorism" means activities that:

> (A) involve acts dangerous to human life that are a violation of the criminal laws of the United States or of any State; (B) appear to be intended- (i) to intimidate or coerce a civilian population; (ii) to influence the policy

---

[37] Available at: http://news.zdnet.com/2100-1009_22-975305.htm (10 Jul 2012).

[38] USA Patriot Act Boosts Government Powers While Cutting Back on Traditional Checks and Balances: An ACLU Legislative Analysis, ACLU Freedom Network (2001) 16. Available at: http://www.aclu.org/congress110101 a.html (27 Jun 2012).

of a government by intimidation or coercion; or (iii) to affect the conduct of a government by mass destruction, assassination, or kidnapping; and (C) occur primarily within the territorial jurisdiction of the United States".

Section 808 of the Patriot Act is said to be one of the vaguest provisions of the Act. It expands the definition of terrorism to crimes "relating to protection of computers". The language of the Act could encompass a wide range of offences unrelated to terrorism, such as the sale of software that fails to perform correctly, posting incorrect or misleading content on web pages, and deceptive internet marketing schemes. The problem is that, while these "offences" may be serious problems, they do not warrant the retraction of the constitutional liberties of individuals. This has led to criticism of the motivation behind the Act. Senator Feingold believes that the Patriot Act "goes into a lot of areas that have nothing to do with terrorism and have a lot to do with the government and the FBI having a wish list of things they want to do".[39]

It is submitted that the present expanded definition of terrorism will act against many innocent non-citizens on the basis of their political beliefs and associations. Non-citizens could also be detained or deported for providing assistance to groups that are not designated as terrorist organisations at all, as long as the activity of the group satisfies the extraordinarily broad definition of terrorism that covers virtually any violent activity. The onus would then be on the non-citizen to prove that his assistance was not intended to further terrorism.[40] The term "international terrorism" means activities that:

(A) involve violent acts or acts dangerous to human life that are a violation of the criminal laws of the United States or of any State, or that would be a criminal violation if committed within the jurisdiction of

[39] S. A. Osher, 'Privacy, computers and the patriot act: the fourth amendment isn't dead, but no one will insure it', *Florida Law Review* (2002) 54(2), p. 541. Available at: http://heinonline.org/HOL. (27 Jun 2012).

[40] USA Patriot Act Boosts Government Powers While Cutting Back on Traditional Checks and Balances: An ACLU Legislative Analysis, ACLU Freedom Network 16 (2001). Available at: http://www.aclu.org/congress110101 a.html (27 Jun 2012).

> the United States or of any State; (B) appear to be intended — (i) to intimidate or coerce a civilian population; (ii) to influence the policy of a government by intimidation or coercion; or (iii) to affect the conduct of a government by mass destruction, assassination, or kidnapping; and (C) occur primarily outside the territorial jurisdiction of the United States, or transcend national boundaries in terms of the means by which they are accomplished, the persons they appear intended to intimidate or coerce, or the locale in which their perpetrators operate or seek asylum.[41]

The definition also encompasses activities that are "dangerous to human life that are a violation of the criminal laws of the United States or of any State" and are intended to "intimidate or coerce a civilian population", "influence the policy of a government by intimidation or coercion", or are undertaken "to affect the conduct of a government by mass destruction, assassination, or kidnapping" while in the jurisdiction of the US. Terrorism is also included in the definition of racketeering.

The definition states "For crimes to be defined as 'terrorist acts' the government must show that they were calculated to influence or affect the conduct of government by intimidation or coercion or to retaliate against government conduct". This means that the definition requires consideration of the mental state of the perpetrator, does not specifically identify the necessary element of violence, and reduces the purpose clause to just achieving political goals (i.e. influencing government conduct).

The definitions of "international" and "domestic terrorism" in Title 18 of the US Code, which were added by Section 802 of the 2001 Patriot Act, although not enumerating specific harms, are still strikingly broad. They provide a basis for prosecuting acts of terrorism by enacting the "limited" purpose of seeking court orders and search warrants against individuals suspected of engaging in terrorist activity. For instance, Section 219 of the Patriot Act gives federal magistrates in any district the power to issue a nationwide search warrant for individuals suspected of domestic or international terrorism. The crime of

---

[41] Section 2331 United States Code.

"acts of terrorism transcending national boundaries" defined in Section 2332(b) Title 18 of the US Code is defined as conduct that occurs across national borders which results in a range of serious harm to persons and property.[42] The transnational terrorism definition is so broad as to have the potential to include cyber attacks directed against the US government. Section 2332(c) provides punishment according to the amount of harm that each act of terrorism causes. The penalty of life imprisonment or death is provided for those acts resulting in death to others. Other acts of terrorism receive lesser maximum punishments, ranging from 10 years for the threat of an attack to 25 years for damage to property, and to 35 years for maiming.

It is feared that the most likely scenario however, is for an act of cyberterrorism to fall under Section 1030 of Title 18, as amended by the "deterrence and prevention of cyber-terrorism" provision in Section 814 of the Patriot Act. Section 1030 is defined as a "federal crime of terrorism" by Section 2332b(g)(5) which is only relevant to the investigating authorities where such an act is "calculated to influence or affect the conduct of government by intimidating or coercing, or to retaliate against government conduct".[43] In so far as it is relevant to this definition of a "federal crime of terrorism", Section 1030 criminalises conduct against protected computer systems as follows:

> (a) Whoever (1) having knowingly accessed a computer without authorization or exceeding authorized access, and by means of such conduct having obtained information that has been determined by the United States Government pursuant to an Executive order or statute to require protection against unauthorized disclosure for reasons of national defence or foreign relations; or (5)(A)(i) knowingly causes the transmission of a program, information, code, or command, and as a result of such conduct, intentionally causes damage without authorization, to a protected computer; [and] (B) by conduct described in clause (i). caused (or, in the case of an attempted offense, would, if completed, have caused) (ii) the modification or impairment, or potential modification or impairment, of the medical examination, diagnosis,

---

[42] 18 USC 2332b (a)–(b).

[43] 18 USC 2332b(g)(5)(A),18 USC 2332b(f).

treatment, or care of 1 or more individuals; (iii) physical injury to any person; (iv) a threat to public health or safety; or (v) damage affecting a computer system used by or for a government entity in furtherance of the administration of justice, national defence, or national security.[44]

However, subsection (c) of Section 1030 provides for a range of independent penalties for the federal crime of interfering with computer systems. As in Section 2332(b), this range of penalties is attuned to the level of harm caused by each cyber attack. Subsection (c) of Section 1030 provides this range of penalties for offences under subsections (a)(i) and (5)(A)(i) as follows:

(c) The punishment for an offense under subsection (a)...of this section is (1)(A) a fine under this title or imprisonment for not more than 10 years, or both, in the case of an offense under subsection (a)(1) of this section which does not occur after a conviction for another offense under this section, or an attempt to commit an offense punishable under this subparagraph; and (B) a fine under this title or imprisonment for not more than 20 years, or both, in the case of an offense under subsection (a)(1) of this section which occurs after a conviction for another offense under this section, or an attempt to commit an offense punishable under this subparagraph[45]; (4)(A) except as provided in paragraph(5), a fine under this title, imprisonment for not more than 10 years, or both, in the case of an offense under subsection (a)(5)(A)(i), or an attempt to commit an offense punishable under that subsection; (C) except as provided in paragraph (5), a fine under this title, imprisonment for not more than 20 years, or both, in the case of an offense under subsection (a)(5)(A)(i) ....or an attempt to commit an offense punishable under [that] subsection, that occurs after a conviction for another offense under this section; and (5)(A) if the offender knowingly or recklessly causes or attempts to cause serious bodily injury from conduct in violation of subsection (a)(5)(A)(i), a fine under this title or imprisonment for not more than 20 years, or both; and (B) if the offender knowingly or recklessly causes or attempts

[44] Section 1030 United States Code.

[45] This subsection was added by the Cyber Security Enhancement Act of 2002.

to cause death from conduct in violation of subsection (a)(5)(A)(i), a fine under this title or imprisonment for any term or years or for life, or both.

Subsection (5)(B) was added by the Cyber Security Enhancement Act of 2002, which was implemented under the Homeland Security Act of 2002. This makes clear that only those offences under subsection (a)(5)(A)(i) (i.e. only those cyber attacks that cause unauthorised damage to a protected computer by inserting code, a program, or other information) which knowingly or recklessly cause or attempt to cause death will attract a maximum penalty of life imprisonment. The Cyber Security Enhancement Act of 2002 directs the US Sentencing Commission to review the sentencing guidelines and ensure that the strength of the penalties reflects the serious nature of computer crime.[46] Cyber attacks under subsection (a)(5)(A)(i) that cause or attempt to cause serious bodily injury will only attract a maximum penalty of 20 years imprisonment, and other breaches of subsections (a)(1) and (a)(5)(A)(i) will attract a maximum of either 10 or 20 years, depending on whether they follow another offence under Section 1030.

The other offences of terrorism which were introduced by the Patriot Act have involved the same approaches by the US government. For example, the offences of destroying an energy facility and destroying national defence materials are broad enough to cover both cyber and physical attacks against infrastructure; however, they only provide a maximum penalty of life imprisonment for attacks that result in death.

Other countries' terrorism- and cyberterrorism-related provisions, such as those of the UK and Malaysia, are broad enough to encompass both cyber and physical attacks against infrastructure; however, they do not make any distinctions between acts causing death, and those causing lesser harm. They provide the maximum penalty of life imprisonment for all politically motivated acts of

---

[46] M. A. Healy, 'How the legal regimes of the European Union and the United States approach Islamic terrorist websites: A comparative analysis', *Tulane Law Review Association* (2009) 84, p. 170.

intimidation (or "influence" in the UK) that interfere with electronic systems and other infrastructure, regardless of an offender's intention to cause a greater level of harm, injury, or death. Although the US was concerned and fearful in the wake of the September 11 attacks, it did not allow this fear to influence its enactments. Unlike the other countries that have enacted sweeping terrorism provisions without any distinction between offences causing death or other injury, the US was very careful to restrict the penalty of death to the most serious acts of terrorism, and calibrate the penalties for attacks causing lesser levels of harm.

It appears that in the terrorism definition, although the behaviour or the action of the perpetrator is significant, the intent of the perpetrator is much more serious. The problem of a definition for terrorism has also given rise to the issue that individual state governments condemn all sorts of violations against countries if they are implemented by any other groups and individuals except themselves.[47]

To sum up, following the September 11 attacks, the US government is developing zero-tolerance policies against terrorism. The military response will not stop the terrorist production line and it is effective in the short term. Western governments work together with the Muslim countries to break the ideology of al-Qaeda and its associated groups by sending the message that violent Islamist groups are not Koranic but heretical. To succeed, the strategy to fight terrorism must become truly multi-pronged, multi-dimensional, multi-agency, multi-jurisdictional, and multi-national.[48]

b. The UK perspective

The broad definition of "terrorism" in the Terrorism Act 2000, provisions proscribing membership in terrorist organisations and the terms of investigative powers, including powers of preventive arrest, have all affected other countries' terrorism definitions. Countries were required to criminalise acts of terrorism under UN Security Council

---

[47] C. Townshend, *Terrorism: A Very Short Introduction*, 1st Edn, Oxford University Press, United Kingdom, 2011, p. 5.

[48] R. Gunaratna, *Inside Al Qaeda: Global Network Of Terror*, 1st Edn, Columbia University Press Publishers, United States of America, 2002, p. 76.

Resolution 1373, but it did not provide a terrorism definition for states to implement in their domestic legislation.[49] In contrast, countries did not come under the influence of the US Patriot Act. The UK Terrorism Act 2000 had much more influence abroad than the US Patriot Act. The definition of terrorism under the Terrorism Act included a definition of 'cyberterrorism'; however, this definition was amended following the rapid changes that have occurred in the cyber world and the emergence of new types of terrorism.[50]

The Terrorism Act 2000 has expanded the notion of terrorism in order to entail a broader definition. It has expanded to encompass forbidden activities and related consequences. This Act was written specifically for Northern Ireland, but its approach was general and it was applicable to other forms of terrorism, because its scope was not confined.[51] The current terrorism definition in the UK, even in its latest version, is still broad. It includes activity that would generally be referred to as terrorism. In the UK Terrorist Act and the Anti-Terrorism Crime and Security Act 2001, the terrorism definition extends to cover non-violent protest. Similarly, Malaysia takes the same position in its terrorism act. It appears that Malaysia has adopted its definition in the Penal Code directly from the UK's definition without any attempt to update or improve on it.

The Terrorism Act 2000 made it illegal to be in a specific terrorist organisation and to support it. The Terrorism Act 2000 added 14 organisations in order to increase the scope of prosecuting terrorism. The membership of a person can be proved even if he committed the act of terrorism from a distance; he is still subject to charges. Put simply, being a member of a proscribed organisation makes the person guilty of an offence and carries a punishment of up to 10 years' imprisonment. Supporting, financing and funding a proscribed organisation were also criminalised under the Terrorism Act 2000.

---

[49] S. Choudhry, *The Migration of Constitution Idea*, 1st Edn, Cambridge University Press, United States of America, 2006, p. 377.

[50] Section 1 Terrorism Act 2000.

[51] M. Charvat, 'A study of UK anti-terror law, legal aspect of combating terrorism', in Centre of Excellence Defence against Terrorism, *Responses to cyber terrorism*, IOS Press, Ankara, Turkey, 2008, p. 106.

Furthering the activities of a proscribed organisation or setting up a meeting with the purpose of encouraging support for a proscribed organisation is also labelled as an act of terrorism under this Act.

Following 11 September 2001, the Anti-Terrorism, Crime and Security Act 2001 was the UK's initial response. This was because; the previous Act (Terrorism Act 2000) was not able to deal with the threat of international terrorism and religious extremism. The later Terrorism Act 2006 criminalised training and encouragement of terrorism, aiding and abetting terrorism and dissemination of terrorist propaganda that could cause others to commit terrorist acts. Preparation, attending terrorist training camps, or purchasing equipment intended to commit a terrorist act were also criminalised according to new Act.[52] It addressed terrorism not only in the UK, but also anywhere in the world. The UK's Terrorism Act 2000 defined terrorism as follows:

> (1) In this Act "terrorism" means the use or threat of action where: (a) the action falls within subsection (2), (b) the use or threat is designed to influence the government or to intimidate the public or a section of the public and (c) the use or threat is made for the purpose of advancing a political, religious or ideological cause. (2) Action falls within this subsection if it: (a) involves serious violence against a person, (b) involves serious damage to property, (c) endangers a person's life, other than that of the person committing the action, (d) creates a serious risk to the health or safety of the public or a section of the public or (e) is designed seriously to interfere with or seriously to disrupt an electronic system.

Section 1(1) of the Act differentiates terrorism from other crimes by requiring that the proscribed harms be "designed to influence the government or to intimidate the public or a section of the public, and the use or threat is made for the purpose of advancing a political, religious or ideological cause". The Terrorism Act 2000 in Section 1 also defines "governments" and "public" to include those from a

[52] M. Charvat, 'A study of UK anti-terror law in legal aspect of combating terrorism', pp. 109–110.

foreign country.[53] A common method used in domestic and international laws to distinguish terrorism from other crimes is the requirement that a terrorist act is used to intimidate the public.

The requirement that terrorists act for the purpose of advancing a political, religious, or ideological cause is less common and more problematic. Political or religious motives go against the traditional criminal law principle that the accused's motive is not an essential element of an offence. Therefore, to prove the essential elements of a terrorism offence, police and prosecutors must collect evidence about a terrorist suspect's religion or politics. The scope of what is meant by a political or ideological cause is also uncertain. The origins of Britain's requirement that terrorism be committed for a political, religious, or ideological motive are interesting. Before 2000, Britain defined terrorism as "the use of violence for political ends, and includes any use of violence for the purpose of putting the public or any section of the public in fear". The reference to the use of violence "for political ends" can be seen as a motive requirement although perhaps a more limited one than the reference to political, ideological, and religious objectives.[54]

Section 1(1)(b) confirms the fact that terrorism causes suffering to governments as well as other bodies, and also causes suffering to government agents such as the police, or to the public. In contrast with the provisions of the Prevention of Terrorism Act 1989, the Terrorism Act 2000 expands on the definition of "terrorism". The previous one was too narrow and gave rise to the situation that it did not catch every single issue and did not cover religious terrorism. Thus, in the Terrorism Act 2000, the definition of "terrorism" was expanded to forbidden activities and related consequences. Generally, Section 1 of the Terrorism Act gives a broader definition and includes legitimate political activity as well as terrorism cases.[55] Furthermore,

---

[53] Section 1 (4)(d) Terrorism Act 2000.

[54] S. Choudhry, *The Migration of Constitution Idea*, 1st Edn, Cambridge University Press, United States of America, 2006, pp. 377–379.

[55] C. Walker, 'Briefing on the Terrorism Act 2000', *Terrorism and Political Violence* (2000) 12(2), p. 5. Available at: http://dx.doi.org/10.1080/09546550008427559 (16 Jan 2011).

the police can take action when they are sure that the alleged action is contrary to the criminal law in other countries.

As a result, the Home Office thought that it confined the application of the definition of terrorism especially when it is relevant to international terrorism.[56] Furthermore, Section 1 of the Terrorism Act drops the objective of Section 20 of the 1989 Act in "putting the public or any section of the public in fear", since this may also result from non-political issues and yet still amount to terrorism as well. The Home Office considers that the predecessor act was focused on "violence"; therefore, it might limit the act to a threat on personal safety. The notion of 'violence' implies unjustified and unlawful force, which usually entails criminal offences involving a threat to, or endangerment of, personal safety.[57]

According to Section 2 of the Terrorism Act 2000, a politically or religiously motivated use or threat of action designed to influence any government or intimidate the public will constitute a terrorist activity if it:

> (a) involves serious violence against a person, (b) involves serious damage to property, (c) endangers a person's life, other than that of the person committing the action, (d) creates a serious risk to the health or safety of the public or a section of the public, or (e) is designed seriously to interfere with or seriously to disrupt an electronic system.

However, it was found that the three conjunctive parts in Section 1(1) were too broad in order to serve as a means for investigative powers. It appears that there is room for error. For example, there have been complaints about the use of stop and search powers under Section 44 of the Terrorism Act 2000 against young Asian males. Since it is too late to change the usage and it will indeed grow with the Terrorism Act 2006, it is submitted that there should be reform in the definition. In terms of the definition in Section 1(1), relevant measures should be taken around a combination of the types of seriously threatening and destabilising offences being perpetrated; and in

---

[56] L. Berwick *et al.*, *Inquiry into Legislation against Terrorism*, Stationary Office Book Publishers, London, United Kingdom, 1996, pp. 85–90.

[57] C. Walker, 'Briefing on the Terrorism Act 2000', p. 5.

terms of context and remit as in Section 1(2), what is important is the presence of collectives that carry them out and that render less capable normal criminal justice processes. In this way, the emphasis should be upon severe and collective political violence, rather than terrorism *per se*.[58]

The Terrorism Act 2000 deviated from the previous focus of 1989 Act, which was violence. This was because, the notion of "violence" implied unjustified and unlawful force, which usually entails criminal offences involving a threat to, or endangerment of personal safety.[59] Consequently, it removes some parts of the act, which might not be relevant, but keeps those parts the impact of which could be devastating, such as the effects on key computer systems or supply of water and electricity. The view of the Home Office, the FBI definition, and that of some critics is that the term violence "would not cover hacking into vital computer installations or contaminating public utilities".[60] Therefore, it is not accurate to say that the terrorism definition is extended by Section 1(2)(b) to cover damage to property. In the modern age, terrorism has more focus on power that relates to finance and security infrastructure and they have shifted their objectives from territories and places to vital powers. In addition, they have changed their targets and policies. To deal with this, Section 1(2) makes an effort to protect against damage to property, persons, life, and infrastructure (including computer systems). Another significant extension was the addition of Section 1(2)(e), which was designed to take account of cyberterrorism, since it entails such activity as DoS attacks, which do not damage any machinery *per se*.

However, at the report stage, the House of Lords replaced the term 'violence' with 'damage' in subsection 2(b), as if it could carry the implication of violence against property only in certain circumstances, such as graffiti, but it would not include serious damage and terrorism. With the replacement of the term 'violence' with 'damage'

---

[58] C. Walker, 'Clamping down on terrorism in the United Kingdom', *Journal of International Criminal Justice* (2006) 4, p. 1146. Available at: http://jicj.oxfordjournals.org/ (22 June 2012).

[59] C. Walker, 'Briefing on the Terrorism Act 2000', p. 5.

[60] C. Walker, 'Briefing on the Terrorism Act 2000', p. 5.

by the House of Lords in Section 2(b), Section 1(2)(e) also was designed to take account of cyberterrorism.[61]

Section 1(2)(e) of the Anti-Terrorism, Crime and Security Act 2001, was passed to take account of cyberterrorism. This Act did not provide protection against terrorism or British cyber terrorists.[62] Section 1 of the Terrorism Act 2000 includes cyberterrorism in its definition of terrorism with the phrase "is designed seriously to interfere with or seriously to disrupt an electronic system".[63]

According to subsection (2)(e), if a cyber attack occurs as a terrorist act, it would fall under this section. Terrorism activity under the Terrorism Act 2000 includes action "outside the United Kingdom with reference to any person or to property wherever situated; reference to the public which includes the public of a country other than the United Kingdom" as well as reference to the government which includes "the government for the UK ... or of a country other than the UK". Furthermore, Section 5 of this Act states that "a reference to action taken for the purpose of terrorism includes a reference to action taken for the benefit of a proscribed organisation". Put simply, any individual engaged in terrorist activity is subject to prosecution under this Act, regardless of the place in the world he is located, or if he is a member of a proscribed terrorist organisation.[64]

The Anti-terrorism, Crime and Security Act 2001 provides a maximum penalty of life imprisonment for those who commit politically motivated actions that are designed to influence any government by interfering with any electronic systems in any jurisdiction. Having stated this broad definition, it increases the possible targets by including "international government organisations". Most countries merely criminalise acts of terrorism directed at governments and civilian

---

[61] C. Walker, 'Cyber terrorism: Legal principle and law in the United Kingdom', *Penn State Law Review* (2006) 110(3), p. 632.

[62] S. Ramage, *Serious Fraud and Current Issues*, 1st Edn, iUniverse Publication, United Kingdom, 2005, p. 175.

[63] J. Clough *et al.*, *Principle of Cyber-crime*, 1st Edn, Cambridge University Press, United Kingdom, 2010, p. 160.

[64] D. M. Jones, *Globalization and the New Terror: The Asia Pacific Dimension*, Edward Elgar Publishing, United States of America, 2004, p. 133.

populations, such as Australia. It criminalises cyber attacks that destroy electronic systems, interfere with national power and water supplies, cause major economic harm, physically injure civilians, or create a national public emergency, as the UK government understandably wants to deter and prosecute these acts.[65]

Furthermore, it requires the intention of the offender but not the *actus reus* of the offender, e.g. interfering with the system. It is higher than the standard of the terrorism offence in terms of the fault element or *mens rea* by requiring the offender to actually intend such interference. This intent must include 'influence', a standard of intimidation is set by requiring that an act merely intends to 'influence' a government or international organisation. Therefore, a DoS attack which is politically motivated falls within this section. Lord Carlile recommended that the terrorism definition in Section (2)(e) The Anti-terrorism, Crime and Security Act 2001 must be retained, since it can cover a broad range of possible consequences that stem from cyber attacks.[66]

Due to the potential great harm caused by acts of cyberterrorism, the UK Parliament and the House of Lords emphasised the great harm that could be caused by cyber attacks to the public by using computer hacking techniques to destroy electronic systems. Recently, the coalition government's first National Security Strategy announced that cyber attacks and terrorism are the government's highest priority, since the identification of cyber attacks and terrorism are a "tier one" risk to national security, among four major risks.[67] Therefore, the government has extended the definition of terrorism to include cyberterrorism offences.

In addition, subsection (2)(e) of the Anti-terrorism, Crime and Security Act, besides criminalising cyber attacks that destroy

---

[65] K. Hardy, 'Cyber-attack against infrastructure in domestic Anti-terror law', *Computer Law and Security Review* (2011) 27, p. 155. Available at: http:// sienecedirect.Com (22 Oct 2011).

[66] K. Hardy, 'Cyber-attack against infrastructure in domestic Anti-terror law', p. 155.

[67] Helen Mulholland, 'Terrorism and cyber-attacks are main threats to UK, national security strategy finds', *Guardian*, 18 Oct 2010. Available at: http://www.guardian.co.uk/politics/2010/oct/18/terrorism-cyber-attacks-national-security-strategy (22 Nov 2011).

electronic systems, interfere with national power and water supplies, cause major economic harm, physically injure civilians, or create a national public emergency, also applies to offenders who merely intend to seriously interfere with or disrupt non-essential electronic systems.

The additional phrase in 1(2)(e) is "serious disruption to computer systems to advance a political, religious, or ideological cause". "Serious disruption" includes cyberterrorism, despite the fact that "costly nuisance" may not be part of cyberterrorism. Ultimately, it could be said that the definition of terrorism in the new version does not fit with the needs of society, due to the broad definition of this term. Furthermore, it does not differentiate between direct attacks and indirect attacks on the individual to clarify the definition.[68]

The Terrorism Act 2006 established new offences of terrorism. The Terrorism Act 2006 criminalised training and encouragement of terrorism, aiding and abetting terrorism, and dissemination of terrorism propaganda that could cause others to commit terrorist acts. It also covered preparation, attending terrorist training camps, or purchasing equipment intended for a terrorist act.[69] It addressed terrorism not only in the UK, but also anywhere in the world. Section 1 of this Act criminalised the acts of publishers whose literature "is likely to be understood by their audience as a direct or indirect encouragement or other inducement to it to the commission, preparation or instigation of acts of terrorism or specified offences".[70] It is not necessary for the publisher to have an intention, or to be reckless. In both circumstances, he/she will be guilty if his/her statement directly or indirectly encouraged or induced the public. In subsection 3, the indirect encouragement of terrorism must include a statement that glorifies the commission or preparation of acts of terrorism. Apart

---

[68] M. Charvat, 'A study of UK Anti-Terror Law', in *Centre of Excellence Defence Against Terrorism. Legal Aspect of Combating Terrorism,* IOS press, Ankara, Turkey, 2008, pp. 109–110.

[69] M. Charvat, 'A study of UK anti-terror law', in *Legal Aspect of Combating Terrorism*, pp. 109–110.

[70] C. Walker, 'Clamping down on terrorism in the United Kingdom', *Journal of International Criminal Justice* (2006) 4, p. 1141. Available at: http://jicj.oxfordjournals.org/ (22 Jun 2012).

from the cyberterrorism definitions in the US Code and related legislation, US scholars provide countless definitions for this term in their literature, ranging from broad definitions to very focused ones.

Although the new definition for terrorism in the UK attempts to comply with global instruments, this definition still goes beyond the current European Union (EU) Directive[71] and the suggestions arrived at from the report of the UN High-Level Panel on Threats, Challenges, and Change. For instance, it includes acts of terrorism by the use or threats of action where it "involves serious damage to property" but does not specify what constitutes "serious".[72] Section 1 of the Terrorism Act 2000 of the UK defines terrorism[73] as follows:

> (1) In this Act "terrorism" means the use or threat of action where — (a) The action falls within subsection (2), (b) The use or threat is designed to influence the government or an international governmental organisation[74] or to intimidate the public or a section of the public, and (c) The use or threat is made for advancing a political, religious, racial[75] or ideological cause.[76]

The UK expanded terrorism definitions in the Terrorism Act of 2000 and its amendments. The definition now encompasses forbidden activities. Legislators have attempted to broaden the discussion of terrorism to increase the scope of terrorism in this law. The amendment of this act was established following the September 11 attacks as an initial response to this incident. In fact, the political and ideological motivations that influence the government have played a pivotal role in this definition. The amendments to the Terrorism Act of the UK have broadened the scope to cover every single issue of terrorism and cyberterrorism. The Anti-Terrorism, Crime and Security

---

[71] EU directive laid down certain rules that must be adapted by national authorities of member states in their laws; however, they are free to decide how to do so.

[72] Available at: http://www.quaker.org.uk/definition-terrorism-uk-law (20 Feb 2012).

[73] Section 1 (1) Terrorism Act 2000.

[74] Inserted (13.4.2006) by Terrorism Act 2006.

[75] Inserted (16.2.2009) by Counter-Terrorism Act 2008.

[76] Section 1 of the Terrorism Act 2000.

Act of 2001 was passed to address cyberterrorism. This law provides the maximum penalty of life imprisonment for those who commit politically motivated actions that are designed to influence any government by interfering with any electronic systems in any jurisdiction.

c. The Malaysian perspective

Malaysia, as an Asian country developing in rapid strides, has attempted to come up with laws relating to terrorism. In Malaysia, amendments have been made to the Penal Code by adding Chapter VIA relating to terrorism in 2004 and the drafting of the Anti-Money Laundering and Anti-Terrorism Financing Act 2001. Chapter VIA Section 130B of the Penal Code describes terrorism. This chapter defines terrorism and related subjects and under 130B(1), it defines related terms to this chapter. It describes "terrorists", "terrorist group" and "terrorist". The next provision, 130B(2), defines what constitutes a "terrorist act". However, the scope of the provision is too general and broad. It includes everything that appears to be related to terrorism. Similar to other legislators around the world who battle such types of threats in their own jurisdictions, the Malaysian Penal Code amendment Chapter VIA (offences relating to terrorism) in 2003 and 2007 include cyberterrorism.[77] However, it is submitted that Chapter VIA needs reconsidering and important safeguards. At present it appears to be formulated to be used as a tool to advance a political agenda and to perpetuate a climate of fear.[78]

Meanwhile, amendments to the Criminal Procedure Code have given the police the authority to commence ancillary investigations on offences relating to terrorism under Chapter XIIA. However, in actual fact Malaysia has always had a specific Act that deals with terrorism or any form of threat to national security, namely the Internal

[77] A. Hana Wahid, 'Offences of the new world — understanding e-crimes' (2010) 1 LNS lii, Available at: http://www.cljlaw.com.www.ezplib.ukm.my/membersentry/articlesdisplayformat.asp? (17 Oct 2011).

[78] Submission of Suara Rakyat Malaysia (SUARAM) to The Eminent Jurists Panel on Terrorism, Counter-Terrorism and Human Rights South East Asia, Suara Rakyat Malaysia, 23 July 2006, p. 7.

Security Act 1960 which was drafted in accordance with Article 149 of the Federal Constitution. However, the Internal Security Act 1960 has been repealed and the government has enacted a new rule modelled on the Patriot Act (2001) of the US and the Anti-Terrorism Act in the UK.

The Internal Security Act has been replaced with the Security Offences (Special Measures) Act 2012. This provides special measures relating to security offences. The Malaysian Penal Code amendments to Chapter VIA (offences relating to terrorism) in 2003 and 2007 stand to include cyberterrorism.[79] Section 130B(1) defines a terrorist entity as any entity that is owned or controlled by any terrorist or terrorist group, including the associations of such an entity. In order to comprehend this definition, one must first understand the meaning of the term "terrorist" itself before one may be able to identify a terrorist entity. However, the term "terrorist" is undefined. It appears as though the approach used in Malaysia is that it would be easier to identify the term "terrorist" by referring to the act of terrorism. In this context, the term "terrorist group" is said to mean:

> "(a) an entity whereby one of its activity or aim is to commit or assist in the commission of terrorists acts; or (b) entity determined under Section 66B or Section 66C of the Anti-Money Laundering and Anti-Terrorism Financing Act 2001".[80]

Section 130B(2) discusses what amounts to an act of terrorism. This subsection states that the act of terrorism is an act or threat either within or outside of Malaysia that falls under Section 130B(3), an act or threat made with the intention to extend a political, religious, or ideological cause and that the said act or threat may be reasonably intended to invoke fear in the public or influence or force the government or international organisation to be compelled to or prohibited from doing a certain act.

---

[79] A. Hana Wahid, 'Offences of the new world understanding e-crimes', LNS lii.

[80] Siti Zaharah *et al.*, *Undang Undang Jenayah Di Malaysia*, 1st Edn, MDC Publishers, Malaysia, 2010, pp. 75–58.

The definition of what constitutes a terrorist act is given from Section 130B(2)(a) to Section 130B(2)(j). As has been said, it is too general and wide, and ranges from:

(a) Serious bodily injury to a person;
(b) Serious damage to property;
(c) Endangering a person's life;
(d) Creating a serious risk to the health or safety of the public;
(e) Involving the use of firearms, explosives and other lethal devices;
(f) Involving releasing into the environment any dangerous, hazardous, radioactive, or harmful substance; toxic chemical; microbial, or other biological agent or toxin;
(g) Disrupting or intending to disrupt or seriously interfering with any computer system related to communications infrastructure, banking, or financial services;
(h) Intending to disrupt the provision of essential emergency services such as police, civil defence, or medical services; and then the catch-all involving prejudice to national security or public safety.
(i) Combinations of the above (a)–(i).

Such clarification on terrorist acts and who constitute terrorists gives more arbitrary powers to the authorities. What constitutes a terrorism offence is open to wide interpretation and may include anything and everything.

In Section 130B of Penal Code, legislators elevated simple offences to terrorism offences solely relying on the loosely drafted definition of "intention". The legislators have applied *mens rea* in the form of intention or knowledge or reason to believe in determining whether an accused should be made responsible for the offence. The standard measurement for intention is ambiguous and vague. The required *mens rea* in Section 130B(2)(aa) and (bb) to warrant prosecution will be "to intimidate the public" or "to influence the government or any international organisation from doing or refraining from doing any act".[81]

[81] Suara Rakyat Malaysia (SUARAM), Submission of Suara Rakyat Malaysia (SUARAM) to The Eminent Jurists Panel on Terrorism, Counter-Terrorism and Human Rights, South East Asia, 23–24 July 2006, p. 7.

The government's intent is to prevent international and state governments from performing any act which is a subspecies to a terrorist act. They did not define 'national security' and 'public safety'; thus, this may lead to misuse of the application of the terrorism definition provision by considering an innocent person being prejudicial to national security or public safety. In addition, any political dissent may be interpreted as a terrorist act under such a vague and broad definition.[82]

For instance, it may affect a wide range of actions which are carried out by non-governmental organisations. Under such a broad definition, the actions of NGOs may be included in this category. Thus, every public protest, demonstration, and peaceful disobedience, as well as political and trade union activities can be labelled as terrorism and may be prosecuted under Section 130 of the Penal Code. Furthermore, these activities can be repressed and be made vulnerable to prosecution, according to this provision.[83]

In July 2004, a committee was set up to review the Penal Code amendments. They recommended a new clause to tighten the "terrorist act" definition with the additional intention "of advancing a political, religious, or ideological cause". Moreover, the committee added an exception clause for acts with the purpose of "advocacy, protest, dissent, or industrial action" and which "is not intended to injure and endanger another person's life". Both recommendations still suffer from imprecision and they are dependent upon the benevolence of judges and the public prosecutor.[84]

Section 140B(3) lists the types of acts or threats that may be considered as acts of terrorism. Paragraphs (a) to (k) of this subsection show the various forms of acts or threats of terrorism. This includes injury or death, damage to property, use of firearms, explosives, ammunition, threats to health and the environment, acts of sabotage

---

[82] Attack on Justice- Malaysia, p. 3.

[83] Suara Rakyat Malaysia (SUARAM), Submission of Suara Rakyat Malaysia (SUARAM) to The Eminent Jurists Panel on Terrorism, p. 8.

[84] Suara Rakyat Malaysia (SUARAM), Submission of Suara Rakyat Malaysia (SUARAM) to The Eminent Jurists Panel on Terrorism, Counter-Terrorism and Human Rights, South East Asia, 23–24 July 2006, p. 8.

towards communication system services, financial services, basic amenities, the country's defences, emergency services, and flight offences. All the acts mentioned in these paragraphs are among the acts that are most often identified with acts of terrorism. In order to constitute a terrorist group, on the other hand, the said entity must have committed one of the said acts or threats in paragraphs (a)–(k) as part of their activity, or assisted in the commission of the same. Since the requirement is that only one of the acts listed in (a)–(k) must have been committed in order for an organisation to be labelled a terrorist group, this section provides a seemingly wide definition. However, this definition is too broad, general and vague.

This definition contains other types of crimes and promotes them to the offence of terrorism to intimidate the public and organisations from committing such acts.[85] Similarly, legislators have elevated simple offences to the terrorism offence solely relying on the loosely drafted definition of intention. Legislators have applied *mens rea* in the form of intention or knowledge or reason to believe in determining whether an accused should be made responsible for the offence. The standard measurement of intention is ambiguous and vague. The required *mens rea* in Section 130B(2)(aa) and (bb) to warrant prosecution is "to intimidate the public" or "to influence the government or any international organization from doing or refraining from doing any act".[86]

By this broad definition, every public protest and demonstration and peaceful disobedience and political and trade union activities can be labelled as terrorism and have sufficient potential to be prosecuted under Section 130 of the Penal Code. Furthermore, such activities can be repressed and made vulnerable to prosecution, according to this provision.[87]

---

[85] SUARAM, Recommendations to the Special Committee on the Criminal Procedure Code and the Penal Code, p. 3.

[86] Suara Rakyat Malaysia (SUARAM), Submission of Suara Rakyat Malaysia (SUARAM) to The Eminent Jurists Panel on Terrorism, Counter-Terrorism and Human Rights, South East Asia, 23–24 July 2006, p. 7.

[87] Suara Rakyat Malaysia (SUARAM), Submission of Suara Rakyat Malaysia (SUARAM) to The Eminent Jurists Panel on Terrorism, Counter-Terrorism and Human Rights, South East Asia, 23–24 July 2006, p. 8.

d. Comparative analysis

This section provides an overview and critique of the legislation that is used to define and categorise acts as terrorism and cyberterrorism. It looks at the enactment of legislation, and considers the requirements for an act to qualify as an act of cyberterrorism in each jurisdiction. It is noted that all three of the countries' definitions have extraordinarily low-harm requirements for an act of terrorism directed against an electronic system or other infrastructure. Of course, a cyber attack in the US, the UK, or Malaysia would need to satisfy other necessary requirements to qualify as a terrorist act. The attack must be politically motivated and must be seriously intended to influence a government or intimidate a civilian population.

The UK and Malaysia's terrorism- and cyberterrorism-related provisions, are broad enough to encompass both cyber and physical attacks against infrastructure; however, they do not make any distinctions between acts causing death, and those causing lesser harm. They provide the maximum penalty of life imprisonment for all politically motivated acts of intimidation (or "influence" in the UK) that interfere with electronic systems and other infrastructure, regardless of an offender's intention to cause a greater level of harm, injury, or death. The current terrorism definition in the UK, even in its latest version, is still broad. It includes activity that would generally be referred to as terrorism. In the UK Terrorist Act and the Anti-Terrorism Crime and Security Act 2001, the terrorism definition extends to cover non-violent protest. Similarly, Malaysia takes the same position in its terrorism act. It appears that Malaysia has adopted its definition in the Penal Code directly from the UK definition without any attempt to update or improve on it. Although the US was concerned and fearful in the wake of the September 11 attacks, it did not allow this fear to influence its enactments. Unlike the other countries that have enacted sweeping terrorism provisions without any distinction between offences causing death or other injury, the US was very careful to restrict the penalty of death to the most serious acts of terrorism, and calibrate the penalties for attacks causing lesser levels of harm.

Most of the definitions related to terrorism and offences related to terrorism in Malaysia are vague and ambiguous, the definitions of terrorism in the Internal Security Act and the Penal Code are not much different. Both definitions are consistent in that they apply to any aspect of violence that concludes in political gain. According to a survey which considered several countries, six common elements may be discovered in the definition of "terrorism" and they are: force, political, fear, terror, threat, and psychological effect.

Due to the mounting pressure within Malaysia for major changes in its anti-terrorism laws, the Security Offence Act, which provided for detention without trial in Malaysia, was repealed in 2012. The Security Bill of 2012 was tabled on 10 April 2012. However, it appears that this new Act will cause more violations of the rights of Malaysian citizens. It gives broad powers to the police to infringe on personal liberty and privacy. Comparing Malaysia with the other two countries (the US and the UK), the Malaysian definition of terrorism is too broad, general, and ambiguous. It refers to other types of crimes and elevates them to the offence of terrorism to intimidate the public and organisations from committing such acts.[88] It does not differentiate among the crimes of terrorism that are enumerated in its definition of terrorism. Taken together, as detailed in the foregoing chapters, each of these three jurisdictions makes its own statutes, additions, and subtractions to such a catch-all definition, but the essence of each legislative regime is the same.

ii. *Definition derived from scholars*

There is no single common definition of the term "terrorism" which has gained universal acceptance. Likewise, no single definition of the term "cyberterrorism" has been universally accepted. Although various definitions have been put forward to define "cyberterrorism", none of them have provided an international definition that has been accepted by all countries. In fact, scholars who are expert in this field define "cyberterrorism" with different foci. However, all these attacks occur in cyber space; therefore, they still exhibit some common elements to all acts of terrorism. Cyber terrorist attacks are premeditated

[88] SUARAM Recommendations to the Special Committee on Criminal Procedure Code and Penal Code, p. 3.

and must be planned since they involve the development or acquisition of software to carry out an attack. Computer terrorism is an act that is intended to corrupt or destroy a computer system.[89] Cyber terrorists are hackers with a political motivation; their attacks can impact political structures through corruption and destruction. "Cyber terrorist attacks often target a civilian interest which qualifies cyberterrorism as an attack that results in violence against persons or property, or at least causes enough harm to generate fear".[90]

However, until now the most acceptable definition is that offered by Dorothy Denning, a Professor at George Town University. She defined:

> Cyberterrorism as the convergence of terrorism and cyber space; it is generally understood to mean unlawful attacks and threats of attack against computers, networks, and the information stored therein when done to intimidate or coerce a government or its people in furtherance of political or social objectives. Further, to qualify as cyberterrorism, an attack should result in violence against persons or property, or at least cause enough harm to generate fear. Attacks that lead to death or bodily injury, explosions, plane crashes, water contamination, or severe economic loss would be examples. Serious attacks against critical infrastructures could be acts of cyberterrorism, depending on their impact. Attacks that disrupt non-essential services or that are mainly a costly nuisance would not.[91]

Thus, cyberterrorism is the use of computer networks in order to harm human life or to sabotage critical national infrastructures in a way that may cause harm to human life. Terrorists use computer technology to advance their primary goals of demoralising civilians and destabilising governments. From the terrorism and cyberterrorism definitions in three countries (the UK, the US, and Malaysia), it is

---

[89] P. Galley, 'Computer terrorism: What are the risks?', (1998). Available at: http://www.genevalink.ch /pgalley/ infosec/stsen/terrinfo.html (19 Jul 2010).

[90] D. E. Denning, 'Cyber terrorism', 2000, p. 2.

[91] D. Denning, Cyber Terrorism: Testimony before the Special Oversight Panel on Terrorism, Committee on Armed Services, U.S. House of Representatives, the Terrorism Research Center, 2000.

clear that cyberterrorism and cyber-based terrorist organisations are regarded in the same light as real world terrorism. At the same time, national statutes and amendments to anti-terrorism laws have increased in order to upgrade the ability to confront the borderless and transnational nature of cyberterrorism as a new form of terrorism.

The first step in understanding what cyberterrorism is, is to analyse how terrorists use computer technology in order to demoralise civilians and thereby undermine the ability of a society to defend itself. Computer technology used for the purpose of cyber attacks can be divided into three categories: weapons of mass destruction, weapons of mass distraction, and weapons of mass disruption.

On the basis of the definition and the challenges which have existed behind the term "terrorism" and consequently, for "cyberterrorism", it is beneficial to divide cyberterrorism generally into two categories. The first and broadest meaning is using the internet as the mode or the object of attack. In this sense, the perpetrator utilises the internet to launch its attack. For instance, the attacker may use a DDoS attack against a government website and use encrypted email to conceal its identity. The second category is using the internet as a way of gathering intelligence or information on weapons or weapons training, or is the situation where the technology is used to facilitate terrorist activity which is called ancillary cyber activities.[92]

From all the discussions of terrorism, it has been found that the very first step in furthering the offence of cyberterrorism is often the illegal accessing of a computer system over the internet which is called unauthorised access (it will be discussed in Chapter IV through legal provision), then using software or deploying some other cyber attack methods to launch the attack. In other words, as soon as access to a computer system is achieved, several ways lie ahead for the attacker, ranging from disruptive attacks to destructive attacks.

Most international or national legal instruments criminalise cyberterrorism in a general way, regardless of the terrorism intent and of the result. However, these criteria as well as the technical methods

[92] J. Clough, *Principle of Cyber-crime*, p. 12.

of attack are crucial from a legal point of view.[93] It is necessary to separate action and motivation in defining cyberterrorism activity. It might be said that the consequences of the acts of hacking and acts of terrorism are the same, but the intentional abuse of IT is a part of the terrorist activity in the legal sense. Finally, from the various definitions offered by different scholars, some common elements have emerged. Apart from the cyberterrorism definitions in the US Code and related legislation, US scholars provide countless definitions for this term in their literature, ranging from broad definitions to very focused ones.

Bearing all above in mind, in addition to federal law, several states in the US have passed legislation about cyber crime. Cyberterrorism crime has also been mentioned in certain statutes, but no governmental definition of cyberterrorism exists. Despite the fact that the term 'cyberterrorism' is neither defined in federal law, nor defined by states law, it is used frequently. Therefore, the definitions offered by scholars are implemented. Some of the definitions of cyberterrorism in the US are as follows.

Dorothy Denning in a Testimony before the Special Oversight Panel on Terrorism, Committee on Armed Services, US House of Representatives defined it as: "generally understood to mean unlawful attacks and threats of attack against computers, networks and the information stored therein when done to intimidate or coerce a government or its people in furtherance of political or social objectives".[94]

Mark Pollitt defined it as "premeditated, politically motivated attack against information, computer systems, computer programs and data which results in violence against non-combatant targets by sub-national groups or clandestine agents".[95] He constructed this definition from the combination of Collin's definition of cyber space

---

[93] P. W. Brunst, 'Legal aspect of cyber terrorism', p. 67.

[94] Cyberterrorism: Testimony before the Special Oversight Panel on Terrorism, Committee on Armed Services, U.S. House of Representatives, 105 Cong. May 23, 2000, testimony of Dorothy E. Denning, Professor, Georgetown University. Available at: http://www.cs.georgetown.edu (10 Jul 2012).

[95] M. M. Pollitt, Cyber terrorism — Fact or Fancy? *Proceedings of the 20th National Information Systems Security Conference*, United States, 1997, p. 285.

and the US Department of State's definition. It appears that Pollitt and Denning's definitions are the best definitions available for cyberterrorism. This is because Pollitt and Denning are the only persons who recognise and make the meaningless term of cyberterrorism explicit by considering the relational elements that the word is composed of. Some authors are eager to adopt a broader definition, but such a broader definition may confuse hacking or cyber crime with cyberterrorism. As has been seen, the cyberterrorism definition that is offered by Pollitt and the State Department includes the two integral components of a terrorism definition — the use of violence or force and a political motivation — that is accepted by 80% of scholars. They did not allow for the inclusion of pure information system abuse and extend existing definitions of terrorism to include the destruction of digital property. Devost, Houghton and Pollard allow for the inclusion of pure information system abuse which allows that the attack does not necessarily have to result in violence against humans to be characterised as terrorism.[96]

Most domestic laws define classical or political terrorism as requiring violence or the threat to or the taking of human life for political or ideological ends. The FBI in the US defines cyberterrorism as "a criminal act perpetrated by the use of computers and telecommunications capabilities, resulting in violence, destruction and/or disruption of services to create fear by causing confusion and uncertainty within a given population, with the goal of influencing a government or population to conform to a particular political, social, or ideological agenda".[97]

The State Department's and FBI's definitions are subsumed by the Department of Defense definition contained in regulation O-2000.12 H (which includes "malicious property destruction" as a type of terrorist attack and "destruction at the level of binary code" as the use of special weapons). The FBI definition of terrorism

[96] M. Conway, 'Cyber terrorism: Hype or Reality', *Computer Fraud and Security* (2007) 2, p. 10.

[97] H. M. Hendershot, Cyber Crime 2003 — Terrorists' Activity in Cyberspace, Briefing slides from the Cyber Division, Federal Bureau of Investigation, Washington. D. C. Available at: http://www.4law.co.il/L373.pdf; Internet (23 Feb 2012).

includes acts against terrorism, while the State Department definition does not identify actions taken against property as terrorism. The scholars who support disruption of digital property, along with other conditions, as constituting terrorism, seem to support a definition "that massively extends the terrorist remit by removing the requirement for violence resulting in death and/or serious destruction from the definition of terrorism and lowering the threshold to 'disruption of property'.[98]

The US allocates more time and energy in the field of cyberterrorism than any other country; it undertakes preparations for all threats of the new millennium. Most of its efforts were initiated following 11 September 2001. A week after the 11 September attacks, the FBI announced an overhaul of its top management in order to place more emphasis on counter-terrorism and cyber-crime. It is not surprising that the US government has stepped up its efforts to promote computer and network security to prevent cyberterrorism attacks. Consequently, Congress has enacted several forms of legislation.

The US Patriot Act (2001) divided the term 'terrorism' into two categories: 'domestic terrorism' and 'international terrorism'.[99] The term 'domestic terrorism' covers "activities that, (A) involve acts dangerous to human life that are a violation of the criminal laws of the United States or of any State; (B) appear to be intended, (i) to intimidate or coerce a civilian population; (ii) to influence the policy of a government by intimidation or coercion; or (iii) to affect the conduct of a government by mass destruction, assassination, or kidnapping; and (C) occur primarily within the territorial jurisdiction of the United States".[100]

A number of authors in the US divide cyberterrorism into two categories: "traditional cyberterrorism" and "pure cyberterrorism". They define traditional cyberterrorism as featuring computers as the

---

[98] M. Conway, Cyber Terrorism: Hype and Reality, Dublin City University, Centre for International Studies. In: A. Leigh (Ed.), *Information warfare: Separating hype from reality*, Potomac Books Inc., Ireland, p. 81.

[99] Section 2332b, Title 18 Patriot Act (2001) United States Code.

[100] Section 18.I.113b United States Code (Act 2331).

target or the tool of the attack, while pure cyberterrorism is more restricted as it is limited to attacks against computers, networks, etc.[101] They believe that the Denning's definition is less comprehensive because it is "limited to issues where we would agree is pure cyberterrorism". They think that according to the Denning's definition "many other factors and abilities of the virtual world are leveraged by terrorists in order to complete his mission".[102]

They identify any act of terrorism that utilises "information systems or computer technology as either a weapon or a target as cyberterrorism". Other authors such as Denning and Nelson include physical attack upon information infrastructure in this category. Nelson identifies two new categories of "cyber terror support" and "terrorist use of the net" instead of including such activities in the definition of cyberterrorism.[103] The Federal Emergency and Management Agency (FEMA) presented the concept of cyberterrorism in its own perspective. An attack qualified as cyberterrorism if it caused violence against property or persons or "at least caused enough harm to generate fear".[104]

The Act, in Section 803, establishes a separate offence which punishes harbouring terrorists by imprisonment for not more than 10 years and/or a fine of not more than $250,000, 18 U.S.C. 2339. Section 411 of Patriot Act redefined two categories of terrorism-related factors from five categories rendering an alien admissible, which were recognised by prior law. It redefined engaging in terrorist activity and representing a terrorist organisation and added espousing terrorist activity, being the spouse or child of an inadmissible alien, associating with a terrorist organisation and intending to engage in activities that could endanger the welfare, safety or security of the US.

---

[101] S. Gordon *et al.*, 'Cyber terrorism?', *Symantec Security Response*, pp. 636–637 and 641.

[102] Sarah Gordon & Richard Ford 'Cyber terrorism?', p. 637.

[103] Nelson *et al.*, 'Cyber terror: Prospects and Implications', p. 10.

[104] S. Ozeren, 'Cyber terrorism and international cooperation: general overview of the available mechanisms to facilitate an overwhelming task', In: Centre of Excellence Defence against Terrorism. *Responses to Cyber Terrorism*, IOS Press Publication, Turkey, 2008, p. 72.

'Terrorist organisations' included groups designated as terrorist organisations under Section 219 of the Immigration and Nationality Act. However, 'terrorist organisation' in Section 411 is an organisation which the Secretary has identified in the Federal Register as having provided material support for, committed, incited, planned, or gathered information on potential targets of, terrorist acts of violence to the definition which is offered by Section 219. Then, the definition was recast to include solicitation on behalf of such organisations, or recruiting on their behalf, or providing them with material support to engage in terrorist activity. If the Attorney General finds any alien associated with a terrorist organisation or intending to engage in conduct related to terrorism that may jeopardise the welfare and security of the US while they are in the country, he/she is inadmissible.

iii. *Definition derived from international conventions*

The struggle of defining terrorism among legal scholars has continued since 1920. They have not gained a consensus on what constitutes terrorism. The approach of these scholars in the international arena has been to adopt a specific model to proscribe specific actions of terrorism. Consequently, during these years, around 17 international conventions have taken place in order to come up with a broad definition of terrorism.[105] These 17 conventions have been dedicated to addressing different aspects of terrorist activities; however, these conventions have focused only on illegalising terrorist activities and have ignored the need to provide a consensus on the term 'terrorism'.

International organisations have yet to reach a unified definition of terrorism. One of the main reasons was that different states had different perspectives on terrorism and they preferred to provide a specific definition about terrorism, as mentioned above.[106] However, this view of a specific definition was refocused in 1999 towards drafting a general definition by the UN International Convention for the

[105] B. Golder, 'What is terrorism? Problem of legal definition', *UNSW Law Journal* (2004) 27(2), pp. 273–275.

[106] B. B. Golder, 'What is terrorism? Problem of legal definition', pp. 273–275.

Suppression of the Financing of Terrorism.[107] The most accepted definition of terrorism is the UN definition in this Convention, since this has been reaffirmed in other international instruments[108] as well.[109]

This definition consists of three main elements. The first element is the physical harm of civilians who are not taking an active part in armed conflict. Second is the intention of the attacker with a designed plan. The third element is that civilians will not know when the next strike will happen and this causes an aftermath threat. In this definition, the identity of the terrorists is not addressed; therefore, it may include individuals, groups, and state entities.[110]

The UN combats terrorism through the Security Council. The efforts of the UN against terrorism are divided into two categories: prior to the September 11 attacks and after them. Before the September 11 attacks, the efforts of the UN in addressing the threat of international terrorism advanced vigorously, particularly by the adoption of Resolution 1373.[111] This Resolution imposed obligations on member states in combating terrorism and established a "counterterrorism committee" to monitor implementation of it. Furthermore, it obliged member states to report on their efforts toward executing Resolution 1373.[112]

Terrorism has been an important part of the agenda of the UN for decades, but after the September 11 attacks, the UN provided a different definition for this term and this is the definition that is now being used. Unlike the relatively specific definition before 11

---

[107] B. Golder, 'What is terrorism? Problem of Legal Definition', pp. 273–275.

[108] It was applied in UN Security Council Resolution 1566.

[109] T. Stephen, 'International criminal law and the response to international terrorism', *New South Wales Law Journal* (2004) 27(2), pp. 461–462.

[110] T. Stephen, 'International criminal law and the response to international terrorism', pp. 461–462.

[111] A. Cohen, 'Cyber terrorism: Are we legally ready?' (2010). United Nations Security Council resolution 1373 adopted on 28 September 2001 unanimously following the 11 September terrorists attacks on the united states. It is a binding resolution, since it was adopted under Chapter VII of the United Nations Charter. The aim of the resolution is to hinder terrorist groups in either ways. It called on all states to adjust their national laws according to international convention on terrorism.

[112] Resolution 1373, United Nations Document S/RES/1373 (28 Sep 2001).

September 2001, the definition after 11 September 2001 provides three general legal elements for an action to be considered as terrorism. It states that "any action constitutes terrorism if it is (1) intended to cause death or serious bodily injury to (2) civilians or non-combatants with (3) the purpose of intimidating a population or compelling a government or an international organization to do or abstain from doing any act".[113]

The UN High-Level Panel on Threats, Challenges and Change in 2004 convened a meeting by the Secretary-General of the UN. It called upon states to set aside their differences in terms of terrorism and adopt the definition that was offered by the Convention on International Terrorism. Secretary-General Kofi Annan endorsed this definition as well. It defined terrorism as "any action, in addition to actions already specified by the existing conventions on aspects of terrorism, the Geneva Conventions and Security Council resolution 1566 (2004), that is intended to cause death or serious bodily harm to civilians or non-combatants, when the purpose of such an act, by its nature or context, is to intimidate a population, or to compel a Government or an international organisation to do or to abstain from doing any act". However, this notion was not applied by member states because they lacked the necessary requirement of criminal law instruments and the Security Council resolutions are not binding.

Some of the definitions offered by international organisations and conventions have disregarded motives in their terrorism definition, such as the OAS Convention to Prevent and Punish Acts of Terrorism that considers terrorism as "common crimes of international significance, regardless of motive".[114] Similarly, the Council of Europe Convention on the Suppression of Terrorism (ECST), although it does not contain a definition, explains terrorism as the purpose of extradition and disregards motives. On the other hand, some other conventions do take note of motive. For example, the US Draft

[113] M. Breakey, 'Cyber terrorism: Origin and impact', p. 2.

[114] Organisation of American States: Convention to Prevent and Punish Acts of Terrorism, Vol. 10, p. 256.

Convention on the Punishment and Prevention of Terrorism Acts considers the motive of terrorism offences as it defines the act of terrorism as an international offence if it is "intended to damage the interests of or obtain concessions from a state or an international organization". Given that terrorism has numerous connotations, ranging from rebellion and civil strife to *coups d'état*, no single definition of terrorism plausibly covers all types of terrorism that have occurred throughout history.[115] The definitions that rely on motives are narrower in scope than those that disregard motives. Some definitions limit the instances of terrorism by only focusing on the motives, whereas others that disregard motives have increased the instances of terrorism because they may treat many offenses as terrorism. Hence, countries and states need to cooperate to fight terrorism. The need for international cooperation has increased since the September 11 attacks.

D. Finding an appropriate definition of cyberterrorism

As mentioned above, the doctrine of cyberterrorism defines it from different aspects and attempts to include all necessary terms. Therefore, in this book, cyberterrorism is divided into two categories (intent-based and effects-based) and different views of cyberterrorism are considered to determine which of them will supply a complete definition. As mentioned earlier, a number of views exist regarding the term "cyberterrorism". Due to these divergent views, the various scholars define the term according to different bases. Some of them define it according to its impact, while others define it according to its result.

i. *Intent-based definition*

"Intent-based cyberterrorism exists when unlawful or politically motivated cyber attacks are done to intimidate or coerce a government or people to further a political objective, or to cause grave harm or

---

[115] O. Yen Nee, *International Responses to Terrorism: The Limits and Possibilities of Legal Control of Terrorism by Regional Arrangement with Particular Reference to Asean*, Institute of Defence and Strategic Studies, Singapore, 2002, p. 7.

severe economic damage".[116] The US FEMA defines cyberterrorism as "unlawful attacks and threats of attack against computers, networks, and the information stored therein when done to intimidate or coerce a government or its people in furtherance of political or social objectives".[117] Other views of cyberterrorism stress the manipulation, modification, and destruction of non-physical items such as data, websites, or the perceptions and attitudes this information can influence. Attacks that destroy electronic records of financial transactions, or permit large-scale electronic theft, would cause significant economic damage to a country, but not truly "exist" in the physical world. Thus, the consequences of a cyberterrorism attack, according to the intent-based definition of cyberterrorism, occurs only in cyber space and does not cause any tangible and physical harm. It occurs mostly in data information and computer systems. For example, changing the information on or appearance of an enemy's official web page allows the terrorist to spread negative perceptions or false information without physical intrusion.

The FBI in the US describes cyberterrorism as:

> a criminal act perpetrated by the use of computers and telecommunications capabilities, resulting in violence, destruction and/or disruption of services to create fear by causing confusion and uncertainty within a given population, with the goal of influencing a government or population to conform to a particular political, social, or ideological agenda.[118]

Security expert Dorothy Denning defines cyberterrorism as the "politically motivated hacking operations intended to cause grave harm

---

[116] Cyber Terrorism Definition, IT Law Wiki. Available at: http://itlaw.wikia.com/wiki/Cyber-terrorism (10 Apr 2010).

[117] C. Wilson, Computer Attack and Cyber Terrorism: Vulnerabilities and Policy Issues for Congress, Order Code RL32114, United States of America CRS Report of Congress 2005, p. 6.

[118] H. M. Hendershot, Cyber Crime 2003 — Terrorists' Activity in Cyberspace, Briefing slides from the Cyber Division, Federal Bureau of Investigation, Washington.D.C. Available at: http://www.4law.co.il /L373.pdf; Internet (23 Feb 2012).

such as loss of life or severe economic damage".[119] Some other experts declare that, "Any deliberate use of IT by terrorist groups and their agents to cause harm constitutes cyberterrorism".[120] According to these definitions, the term 'cyberterrorism' is considered a criminal act. Put simply, the misuse of cyber space is criminalised.[121] Cyberterrorism criminal acts occur on an international scale and are sanctioned in the same way as criminal offences. It is a form of unlawful attack against networks, computers, and information of governments.

Another definition by Kevin Coleman, former chief strategist at Netscape and who writes a Homeland Security-focused column for Directions magazine, is "The premeditated use of disruptive activities, or the threat thereof, against computers and/or networks, with the intention to cause harm or further social, ideological, religious, political or similar objectives or to intimidate any person in furtherance of such objectives".[122] In this definition, the bad faith of the perpetrator is substantial in defining cyberterrorism, and it is based on the intent-based definition. Therefore, he focuses on the *mens rea* which creates the essential part of the crime. The result of the crime is not necessary to create the crime. Nevertheless, the criminal intent or *mens rea* of the perpetrator is the material element of the crime and it is created merely by implementing it.

Most of the experts appear to prefer intent-based attacks as qualifying as cyberterrorism, rather than effects-based attacks. The US Department of Homeland Security (DHS), within the National Infrastructure Protection Center (NIPC), defines cyberterrorism as "a criminal act conducted with computers and resulting in violence, destruction, or death of its targets in an effort to produce terror with

---

[119] Activism, hacktivism, and cyberterrorism: The Internet as a tool for influencing foreign policy. In: J. Arquilla & D. Ronfeldt (Ed.), *Networks and Netwars: The Future of Terror, Crime, and Militancy*, p. 241.
[120] S. Krasavin, 'What is Cyber terrorism?', (2004) Computer Crime Research Center. Available at: http://www.crime -research.org/analytics/Krasavin/ (10 Apr 2001).
[121] Counter-Terrorism Task Force Council of Europe, *Cyber Terrorism: The Use the Internet for Terrorist Purposes*, Council of Europe Publication, France, 2007, p. 144.
[122] K. Coleman, 'Cyber Terrorism' (2003) *Directions Magazine*. Jane Elliott, Publisher. Available at: http://www.directionsmag.com/article php?article (15 Mar 2010).

the purpose of coercing a government to alter its policies".[123] In addition, it includes attacks on computer networks and transmission lines within that definition.

In short, any terrorist act which is conducted in or by means of cyber space or the internet can be called cyberterrorism. However, this definition is not a comprehensive and preventive one; it has two aspects of incongruence. Firstly, it is necessarily broad and includes everything from basic hacking and DoS attacks to concerted efforts to unleash weapons of mass distraction or mass disruption; secondly, such a definition is "limited in application regarding the actor or actors and the intent behind the attack".[124]

Despite its broad definition, this kind of definition of cyberterrorism is limited in application, because it relies on two elements. First, the act of terrorism must be undertaken by an individual, group, or organisation. Therefore, it is separated from similar acts done by either a state or its agents, which might be considered as a use of force under international law. Second, the act of terrorism undertaken by terrorists with the goal of destruction or disruption for political or religious purposes is the secondary goal of cyber terrorists. In terms of the terrorists' goals, cyberterrorism is separate from cyber crime. The key factor is the intent behind the attack and this distinguishes cyberterrorism from related cyber crime.[125]

ii. *Effect-based definition*

Some scholars believe in the strict definition which is confined to attacks resulting in serious harm to persons or property, such as Professor Gabriel Weimann who defines cyberterrorism as "the use of computer network tools to harm or shut down critical national

---

[123] Available at: http://computer.yourdictionary.com/cyberterrorism (7 Sept 2011).

[124] K. Gable, 'Cyber-apocalypse now: Securing the internet against cyber terrorism and using universal jurisdiction as a deterrent', *Journal of Transnational Law* (2010) 43, p. 59.

[125] K. Gable, 'Cyber-apocalypse now: Securing the internet against cyber terrorism and using universal jurisdiction as a deterrent', p. 59.

infrastructures". Effects-based cyberterrorism "exists when computer attacks result in effects that are disruptive enough to generate fear comparable to a traditional act of terrorism, even if done by criminals".[126]

Some security experts define cyberterrorism based on the effects of an attack. For example, attacks where computers are targeted and the resulting effects are destructive or disruptive enough to generate fear that is potentially comparable to that from a traditional act of terrorism, even if not initiated by criminals with any political motive. Under this "effects" view, even computer attacks that are limited in scope, but lead to death, injury, extended power outages, airplane crashes, water contamination, or major loss of confidence in portions of the economy, are defined as cyberterrorism.[127]

A physical attack disrupts the reliability of computer equipment and availability of data. The physical attack focuses on the physical destruction of information hardware and software, or physical damage to personnel or equipment using IT as the medium. A physical attack is implemented either through use of conventional weapons, creating heat, blast, and fragmentation, or through direct manipulation of wiring or equipment, usually after gaining unauthorised physical access.

Examples of this approach would include the chaos and destruction caused by disrupting a nation's air traffic control system, crashing two trains together by overriding the railroad signal and switching system, interfering with the control systems for water or electricity, or blocking and falsifying commercial communications to cause economic disruption.[128] A terrorist group could crash a network through physical destruction or technological attack, but only a group whose perceived gains would offset their loss of information, communication, and other capabilities would do this. Outside

---

[126] Available at: http://itlaw.wikia.com/wiki/Cyber-terrorism (10 Feb 2010).

[127] D. Denning, 'Is Cyber War Next?', (2001) Social Science Research Council, Available at: http://www.ssrc.org/sept11/essays/denning.htm (7 Sept 2011).

[128] US Army Training and Doctrine Command, *Hand Book of Cyber Terrorism and Cyber Operations*, p. 17.

of computer networks, communications networks can also be targeted for destruction, disruption, or hijacking. This has a direct impact on the military and the government since a large percentage of the GIG is dependent on commercial telephone links and the internet. Destructive and disruptive attacks upon communication networks would likely be supporting operations designed to increase the effectiveness of physical attacks. Hijacking or taking control of a communications network might support another operation, or be attempted for its own impact.[129]

In 1991, during Operation Desert Storm, the US military reportedly disrupted Iraqi communications and computer centres by sending cruise missiles to scatter carbon filaments that short-circuited power supply lines. In addition, the al-Qaeda attacks on 11 September 2001, destroyed many important computer databases and disrupted civilian and military financial and communications systems that were globally linked. The temporary loss of communications links and important data added to the effects of the physical attack by closing financial markets for up to a week.[130]

The term "cyberterrorism" differs from other developments of terrorism technology, due to the fact that it includes offensive IT capabilities, either alone or in combination with other forms of attack.[131] Terrorist use of computers as a facilitator of their activities, whether for propaganda, recruitment, data mining, communication, or other purposes, is not cyberterrorism.[132] The law, when including cyberterrorism, must be confined to attacks that result in serious harm to persons or property. Some scholars believe in providing a proper legal response considering a wider notion. As a result, this will include cyberterrorism as well as the various ways the internet is used to sustain terrorism.

---

[129] US Army Training and Doctrine Command, *Hand Book of Cyber Terrorism and Cyber Operations*, p. 17.

[130] L. Carlos *et al.*, 'Cyber Terrorism — A rising threat in the western hemisphere', p. 14.

[131] US Army Training and Doctrine Command, p. 17.

[132] P. Rev, 'Cyber-terrorism: Legal principle and law in the United Kingdom', *Penn State Law Review*, (2006) 113(3), p. 20.

1.3 *MODUS OPERANDI* OF CYBER ATTACK TERRORISM

As discussed, cyberterrorism is a new aspect of terrorism and it differs from other developments in terrorist tools. It involves IT and allows perpetrators to conduct their operations without any physical risk or harm to themselves. They use networks as a medium through which an attack is launched.[133] One of the problems in defining the term "cyberterrorism" and deterring the threat is that it can be carried out in a variety of forms, which leads to the difficulty of defining and stopping them.

It has been found that the very first step in a cyberterrorism offence is often the illegal accessing of a computer system over the internet which is called unauthorised access (unauthorised access will be defined in Chapter IV), then using software or deploying some other cyber attack methods to launch the attack. In other words, as soon as access to a computer system is achieved, several ways lie ahead for the attacker, ranging from disruptive attacks to destructive attacks.

Terrorists launch their attacks by using various methods. The result of some attacks do not only remain in the virtual world, but also lead to real-life harm by the destruction of property or the loss of life. Often, the attack on computer systems seems to be less dangerous than a conventional terrorist attack. Nevertheless, the examples below illustrate that attacks on computer Supervisory Control and Data Acquisition (SCADA) systems, which are used for controlling and measuring other systems (critical infrastructure), can be more effective than conventional attacks.[134] In the virtual world, a computer usually works as a medium and a computer or other computers are viewed as targets for cyber criminals and cyber terrorists.

Cyber terrorists utilise modern high-tech forms to commit their cyber attacks which include Botnet attacks, digital piracy, malicious spreading of viruses, attacks on critical information infrastructure by criminal groups

---

[133] J. Arquilla *et al.*, *Networks and Net Wars*, Santa Monica, RAND Publication, 2001, p. 5.

[134] P. Brunst, Legal Aspects of Cyber Terrorism, p. 65.

and hacking.[135] Other forms are DoS attacks, email spoofing, Internet Service Provider address spoofing, key loggers, logic bombs, sniffers, Trojan horses, viruses, worms, and zombies. However, the main types of attacks are espionage, DoS attacks, logic bombs, and Trojan horses. It is important to know that each type of attack has little to do with the possible impact of the attack. Thus, the type of attack is distinguished and analysed based upon the impact of the cyber attack or intent of the cyber attacker, rather than specific technical means deployed.[136]

Terrorists' first and foremost action is to gain access to a computer system (unauthorised access), and then several ways lie ahead of them. They can manipulate or alter the information therein by shutting down the computer. Within short moments of the administrator shutting down and the restarting the computer system, the attackers can launch their attack. Therefore, even a very short interruption can produce a hazardous situation that terrorists can exploit. Furthermore, if the affected systems have control over power plants or infrastructure, this will provide the best opportunities for terrorists to exploit an outage.[137] They can disrupt or interfere with gas and oil systems, electrical systems, and water supply systems. They can launch their attack via communications and information systems. For example, they can gain access to a drug manufacturer and alter the formula of a drug to a deadly one, or change a patient's blood type by accessing his hospital records.[138]

1.3.1 Classification of Cyberterrorism

Cyber terrorists utilise diverse tools and methods to launch their terrorist program, in particular, hacking (which is facilitated by many

---

[135] C. Ernest, Cyber-crime: New Threat and Global Response, Expert Group on Cyber-crime, Department on New Challenges and Threats, 17–21 January 2011, pp. 15–17.
[136] C. Wilson, Computer Attack and Cyberterrorism: Vulnerabilities and Policy Issues for Congress, Order Code RL32114 CRS Report for Congress, Congressional Research Service Report for Congress, The Library of Congress, 2005, pp. 20–25.
[137] P. Brunst, Legal Aspects of Cyber Terrorism, p. 64.
[138] C. B. Foltz, 'Cyber terrorism, computer crime, and reality', *Information Management and Computer Systems* (2004) 12(2), p. 158. Available at: www.emeraldinsight.com/0968-5227.htm (1 Mar 2012).

technologies, such as packet sniffing, tempest attack, password cracking, and buffering), Trojans, computer viruses, computer worms, email-related crime, DoS attacks, and cryptography. All methods of cyberterrorism include unauthorised access.[139]

A. Botnet attacks

Additionally, they can launch their attack by using Botnets on a large scale. These "zombies" are controlled by command and control servers. Once terrorists have gained unauthorised access to computer systems, they can alter information, modify programs, obtain passwords, and monitor what information is being stored. They can execute these cyber-crime techniques swiftly for cyberterrorism attacks to threaten vast numbers of potential targets, as well as introduce malicious codes, thereby interrupting operations on a global scale. It is submitted that the best method in countering cyber attacks is for nations to accede to an international instrument (international treaty) which most nations accept.

B. Virus and worms attack

Generally, current cyber attacks comprise primarily virus and worm attacks, DoS, web defacement of information sites, and unauthorised intrusion into systems. Web defacement can be done in a second, since it is done through an exploitation tool that searches for vulnerabilities in web servers and makes a list of those vulnerabilities, so that the attacker can then choose his desired web site and desired changes, and the program does the rest. Most of the time, the design and maintenance of web pages are done by people with limited IT skills because; setting up a web page can be done easily through tools such as the Microsoft Web page design tool. A very dangerous defacement can occur when the web page is changed in such a way as to pretend to be part of someone else's system, and then an innocent user submits his or her information. Such an attack, using web pages of large companies or banks, through the use of a popup browser

---

[139] R. Nagpal, Cyber Terrorism. In the Context of Globalization, *II World Congress on Informatics and Law*, Spain, 2002, pp. 4–11.

window that can take a client's login name and password, is called phishing.[140]

Viruses and worms are other types of attacks that infect computers. The difference between a virus and a worm is rather unclear. Both infect computers by being copied and performing a programmed function. A virus infects another program, whereas worms are self-replicating and do not need to infect another application.[141] A good example of a virus used in cyberterrorism is "Stuxnet", which was identified in 2010 and used against the SCADA system in Iran.[142] Scholars believe that the "Stuxnet" malware was created to target and stifle Iran's nuclear system in particular.

There have been unclassified documents that infer that the 'Stuxnet' worm was created specifically to seek out and exploit vulnerabilities in the software that manage the Integrated Computer Solutions(ICSs) found in critical infrastructure facilities.[143] A SCADA system is a type of ICS that controls industrial processes and critical infrastructures. These systems are accessed and managed directly by computer terminals, either remotely or from mobile wireless devices. In 2009, the US DHS conducted an experiment called the Aurora Project that showed that the SCADA systems that control power generators and grids had vulnerabilities to cyber attack present in them. The Aurora Project simulated a computer-based attack on a power generator's control system and caused operations to cease. These same vulnerabilities may exist in other critical infrastructure and if exploited, could both critically affect the economy and have physical consequences, even loss of life.[144]

---

[140] L. Janczewski *et al.*, *Managerial Guide for Handling Cyber Terrorism and Information Warfare*, IDEA Group Publishing, United States, pp. 99–103.

[141] J. Clough, *Principles of Cybercrime*, 1st Edn, Cambridge University Press, 2010, p. 33.

[142] It will be discussed completely in the next pages under "Impact of Cyber Terrorism".

[143] Available at: http://www.ncs.gov/library/tech_bulletins/2004/tib_04-1.pdf (17 Jul 2012).

[144] 'Challenges Remain in DHS' Efforts to Security Control Systems', Department of Homeland Security, Office of Inspector General, August 2009.

The Stuxnet malware infected a particular part of a SCADA system, the programmable logic controller (PLC), and executed its destructive and masking capacities against a PLC with a specific fingerprint, such as the one at Natanz in Iran.[145] Gholam-Reza Jalali, head of the Iranian Passive Defense Organization, said that, "The research and inquiry into the matter indicates that the Stuxnet worm was disseminated from sources in the US and Israel", as well as that "Iran's Foreign Ministry should probe into the political and legal aspects of the cyber-attack while other Iranian bodies should pursue and complain to international circles ... Siemens should explain why and how it provided the enemies with the codes for the SCADA software".[146]

In this case, Stuxnet is considered under the legal international view. The Stuxnet case was neither a simple espionage case nor cyber exploitation; rather, it essentially degraded and disrupted certain SCADA systems. Simply put, the intent of the terrorist organisation was to carry out an attack on critical infrastructure and SCADA systems. Due to the availability of Stuxnet's code, an internet software developer can revise and formulate existing code and terrorist organisations can use them in future.[147] Some scholars believe that the Stuxnet attack was authorised by the UN Charter as a form of self-defence, according to Article 41 which was invoked by the UN. It relies on anticipating self-defence in light of Iran's stated nuclear ambitions and anti-Israel policies and programs. It can be a use of force, but this is in violation of the prohibition in Article 41. Meanwhile, there is considerable debate on this case: was the Stuxnet case really a use of force in violation on Article 2(4) of the UN Charter, or an "armed attack" which allowed a state to claim the right of self-defence under Article 51? Scholars have stated that Stuxnet

[145] Available at: http://opiniojuris.org/2011/01/25/could-deploying-stuxnet-be-a-war-crime/ (26 Apr 2011).

[146] 'Iran May Sue Siemens over Stuxnet Virus'. Available at: https://www.infosecisland.com/blogview/13096-Iran-May-Sue-Siemens-over-Stuxnet-Virus.html (26 Apr 2011).

[147] P. Kerr *et al.*, The Stuxnet Computer Worm: Harbinger of an Emerging Warfare Capability, Congressional Research Service Report for Congress, R41524, p. 3.

does not meet the criteria in the definition of "force" and lacks the physical characteristics. If it is defined using the classical approach, it needs an instrument for its application.[148]

On the other hand, some experts such as Michael Schmitt have suggested that it was a use of force since it was destructive and functioned in the same way as a bomb. Furthermore, considering the definition which was offered by Gray Sharp on "use of force", it qualified as a use of force according to Article 41 which determines "measures not involving the use of armed forced" to include "complete or partial interruption of ... telegraphic, radio, and other means of communication". This is why Stuxnet seems to constitute a use of force: its target was infrastructure, which had critical worth for Iran. Therefore, these suggestions coincide: some experts think it was destructive and could be a use of force, while others think the destructive side was much more selective than a traditional bomb.[149]

Another worm attack called Nimda attacked US cyber infrastructure. It happened seven days after the September 11 attacks. It carried five different malicious payloads via email and automatically infected every computer it met. This worm did not exploit by planting new programs in computers; instead, it immediately exploited the cyber infrastructure by discovering its vulnerabilities. It invaded the financial industry database and computer network. It demonstrated various cyber warfare methods by different approaches such as email, internet, and computer network attacks. It had the ability to self-replicate and find new "doorways" in the internet to other vulnerable computers. It scanned systems for over 100 vulnerabilities and exploited open doors left behind in computers infected with the Code Red worm.

The attack became a global crisis within 30 minutes of being discovered. Richard Clarke, former Chief Counterterrorism Advisor of the National Security Council, described the events on the day of the attack as follows, "Nimda was a devastating attack ... the cyber security team came to me and said there was a major worm going through

---

[148] Available at: http:/opiniojuris.org/2011/01/25/could-deploying-stuxnet-be-a-war-crime/ (27 Apr 2011).

[149] D. Hollis, 'Could Deploying Stuxnet be a War Crime?'. Available at: http://opiniojuris.org/2011/01/25/could-deploying-stuxnet-be-a-war-crime/ (27 Apr 2011).

the internet and it was knocking off major companies". Mr. Clarke said, "We still don't know for sure, but had Nimda happened prior to the September 11 attacks, it would have been a big news story. Many companies, particularly in the financial world, shut down major pieces of their operations. It destroyed and corrupted databases. It was quite devastating, causing several billion dollars in damage".[150] One of the administrators said that the worm infected 50,000–100,000 files at his company's data centre.

Two years after the Nimda attack, the Slammer worm appeared on the scene as the fastest and most destructive attack ever. This worm spread across the globe, in timespans ranging from 1 to 3 minutes, and similar to Nimda, exploited existing Microsoft software vulnerabilities. In the history of worms, Slammer is known as the fastest. It caused disruption to systems in banking, airlines, infrastructure, and emergency services.

Titan Rain was a Chinese-based cyber attack, and invaded military networks of the US at midnight on 1 November 2005. It hit the army information system engineering command at the Defense Information Systems Agency in Arlington, Virginia, and the Naval Ocean Systems Center in San Diego, California. Fortunately, the attack was stopped at 4:46 a.m., as soon as the same vulnerability was found at the US Army Space and Missile Defense installation in Huntsville, Alabama. All unclassified computers and computers that were connected to the internet directly were affected by this attack. But classified computers were not affected by this attack, mainly because classified computers usually have no direct internet connections. It appeared that the attack originated from China.[151] According to the Congressional Report Service (CRS) Report of Congress in 2003, US infrastructure systems are resilient and have the capability to recover easily from cyber attacks. However, security experts have stated that "because technology continuously evolves, it is incorrect to think that future cyber-attacks will always resemble the past annoyances we have expe-

[150] D. Verton, *Black Ice; The Invisible Threat of Cyber Terrorism*, McGraw-Hill, Ventura Publisher, United States of America, 2003, p. 160.

[151] A. M. Half, 'Cyber Power as a Coercive Instrument', Thesis of Advance Air and Space, Maxwell Air Force Base, Alabama, June 2009, pp. 19–20.

rienced from internet hackers".[152] Following the September 11 attacks, investigation revealed that terrorists used email to coordinate detailed instructions about their attack. They exchanged information on the targets and the number of attacks via email.

C. DoS and DDoS attacks

DoS and DDoS attacks have become a serious problem in the current decade. They cause interruptions to services and create irregular false signals in feedback control mechanisms that lead to significant threats to internet services.[153] Permanent elimination of DoS attacks cannot be attained yet, despite the universal transformation of the internet.[154] Thus, conceiving a mechanism to mitigate the impact of DoS attacks is the best avenue so far. One of the best methods to detect a DoS attack is by monitoring the central processing unit (CPU) of a computer. A common way to identify a DoS attack is the high number of packets and high utilisation rate of the CPU that happens during an attack.

The severity of a DoS attack is identified by monitoring the CPU utilisation rate and comparing it with a previous base line of CPU utilisation rate. When a system is targeted by a DoS attack, the service speed of the system slows down and suddenly the number of spam emails increases dramatically. Sometimes, it makes the site unavailable and impossible to access. There are many types of DoS attacks, which are either so disruptive that they prevent users from using the network services, or degrade the quality of service so as to slow down the service. Detecting a DoS attack is a very difficult and confusing job because the detector must distinguish between genuine and bogus data packets in order to determine the organisation that launched an attack. However, no perfect technique for this has been developed so far. To investigate a DoS attack, the first step is to identify the domain name server to trace the internet protocol address by using domain name

---

[152] CRS Report of Congress.

[153] Y. Tang *et al.*, 'Protecting Internet Services from Low-Rate DoS Attacks, In: E. Goetz and S. Shenoi (Eds.), *Critical Infrastructure Protection*, Springer Publication, Boston, 2008, pp. 251–265 and 263.

[154] Wu Zhijun *et al.*, 'An Approach of Defending against DDoS Attack', *Journal of Electronics* (China) (2006) 23(1), pp. 148–153, 153.

server Domain Name System ()logs, and then the investigator can identify the various attacks originating from that attacker.[155]

Nowadays, terrorists also use the internet for exploiting new possibilities in the "war of ideas". They can spread propaganda and use it to advertise their ideas, or send threats to their enemies. Moreover, terrorist use the internet for "harmless" tasks, such as sending email or visiting websites. This allows terrorists to exchange information in a fast-paced and anonymous way.[156]

D. Phlashing

Phlashing is a kind of DoS attack that damages a system badly enough that it forces users to replace or reinstall hardware. It is also known as permanent denial of service (PDoS). It exploits security flaws that allow remote administration on the management interfaces of the victim's hardware, such as printers and routers.[157] The available vulnerabilities are used by the attacker in order to replace a device with a modified or defective firmware which leads in bricking the device and making it unusable.[158]

1.3.2 *Modus Operandi* Adopted by Al-Qaeda

Al-Qaeda unlike many other terrorist organisations, set new types of *modus operandi*, making it all the more formidable. Osama bin Laden encourage his followers to mix various approaches by employing different tactics and operational style. Al-Qaeda was collecting intelligence on targets and sending encrypted messages via the internet. They use internet-based phone services to communicate with cells overseas. Such

---

[155] EC-Council, *Computer Forensics, Investigating Network Intrusion and Cybercrime*, 1st Edn, EC-Council Press, United States of America, pp. 5–9.

[156] P. Brunst, Legal Aspects of Cyber Terrorism, p. 66.

[157] K. Manasa *et al.*, 'Enhanced IDS techniques against attack in wireless mobile Ad-Hoc network', *International Journal of Advanced Research and Innovation* (2013) 2(2), p. 320.

[158] T. Siva *et al.*, 'Controlling various network based ADoS Attacks in cloud computing environment: By using Port Hopping Technique', *International Journal of Engineering and Technology* (2013) 4(5), p. 2101.

usage indicates that the internet is being used as cyber planning tool for terrorists by providing anonymity, command, and control resources and a host of other measures to deploy attack options. This is the most important terrorist tool to attack against information. There are many websites linked to al-Qaeda which contains elements of cyber planning:[159] alneda.com, assam.com, almuhrajiroun.com, qassam.net, jihadunspun.net, 7hj.7hj.com, aloswa.org, drasat.com, jehad.net, mwhoob.net, and aljehad.online.

Several points attract terrorist methodologies to the internet. Terrorists can put together profiles; it enables them to construct a profile to counter their action. The internet creates a true ideological weapon for terrorist, because they can control all broadcasted information anonymously by hiding their identities. Such anonymity is available for either people or internet service providers (ISPs) when the former unwittingly participate in serving people or groups for purposes other than legitimate ones. The web site of www.alneda.com which was for al-Qaeda originally located in Malaysia until 13 May 2002. It reappeared in Texas at http://66.34.191.223/ until 13 June, and then reappeared on 21 June at www.drasat.com in Michigan. It was shut down on 25 June 2002. The ISPs hosting it apparently knew nothing about the content of the site or even the fact that it was housed on their servers. This shell game with their website enabled the al-Qaeda web to remain functional in spite of repeated efforts to shut it down. Cyber deception campaigns will remain a problem for law enforcement personnel for years to come. The internet also provides outstanding command and control mechanisms for terrorists.[160]

1.3.3 Impact of a Cyber Terrorist Attack

The cyber attack can be divided into two categories: (i) attack on an IT system (non-physical destruction) and (ii) attack through the use

[159] T. L. Thomas, 'Al Qaeda and the Internet: The danger of "Cyberplanning"', *US Army War College Quarterly* (2003) 33(1), p. 113.

[160] T. L. Thomas, "Al Qaeda and the Internet: The danger of 'Cyberplanning'", p. 115.

of an IT system (physical destruction). The results of a cyber attack will be either generated on the IT system itself or by using that IT system for destructive purposes. The destruction and disruption that occurs in a cyberterrorism attack focuses on both physical and non-physical destruction.

Physical destruction focuses on destruction of information hardware and software or damage to items using IT as a medium. Terrorists target critical infrastructure, they gain access to these systems, then alter, destroy, modify, and manipulate them. They create chaos and disrupt critical infrastructures, such as air traffic control systems, water systems, electrical systems, energy systems, emergency services, and banking and finance systems.[161] This kind of destruction in cyber attacks results in physical harm or destruction through an IT system and the effects will be caused by physical destruction.

SCADA systems control critical infrastructures — such as electrical and nuclear power systems, telecommunications, and oil storage facilities — and production networks. SCADA systems can be attacked, causing their operations to malfunction. For instance, terrorists may take control of an air traffic control system and cause aircraft to crash. A case occurred in Australia in 2001 where an individual took over a control system and used a wireless radio to cause the release of up to 1 million litres of sewage into the Queensland River.[162] Another case was 'Stuxnet' which was a powerful virus that wreaked havoc with SCADA systems and it specifically attacked Siemens PLCs used to control enrichment centrifuges. It was first identified in 2010.

German researchers with network time announced that the malicious code was traced back to the US and Israel. "The Stuxnet virus attacks are thought to have caused severe damage to Iranian uranium enrichment facilities and reportedly set back the nation's nuclear

---

[161] US Army Training and Doctrine Command, *Hand Book of Cyber Terrorism and Cyber Operations*, pp. 18–44.

[162] R. T. Lemos, 'What are the real risks of cyber terrorism?' (2002), [Online magazine] *ZDNet*, Available at: http://zdnet.com.com/2102-1105_2-955293.html (23 Feb 2012).

program by as much as several years".[163] It was a piece of malware that consisted of multiple parts. It did not rely on one zero-day vulnerability and used at least four. The utilisation of one zero-day dated back to 2008. Although Microsoft patched the flaw, it had not installed it for many systems and process control systems. Therefore, they were wide open to the attack, due to the gap existing in the security of the process control systems. The writers of Stuxnet did not opt for multilayer encryption; thus, the developers stole legitimate certificates from still-active enterprises Realtek Semiconductor and JMicron.[164] It attacked a target group of computers instead of infecting many computers. In doing so, the malware exploited infection vectors that required physical contact with infected devices such as spreading via USB sticks and devices that supported them, like scanners. Consequently, when the initial infections occurred, the malware infected only three PCs in order to remain unnoticed. Following the installation, Stuxnet scanned the system to see "if the new host was up to date and capable of quickly discovering malware activity, and it also checked if the SCADA system was installed".[165] If the malware found a SCADA system, it reprogrammed the PLC. Once it was executed, the attacker had complete control over the production.

Scholars have stated that it appeared that the Stuxnet malware was created in particular to target and stifle Iran's nuclear system. There are indications that the main target of Stuxnet was the Iranian nuclear industry. Stuxnet, unlike other malware, was not designed to steal information, but rather to target and disrupt control systems and disable operations. It has affected numerous countries in varying degrees of disruption to their technology systems. The amounts of damage caused by Stuxnet in such countries is not publicly available and makes it difficult to determine the malware's potency.[166]

---

[163] 'Iran May Sue Siemens over Stuxnet Virus', https://www.infosecisland.com/blogview/13096-Iran-May-Sue-Siemens-over-Stuxnet-Virus.html (26 Apr 2011).

[164] E. Willems *et al.*, 'Cyber Terrorism in the Process of Industry', *Computer Fraud and Security* (2011) 3, p. 17.

[165] E. Willems *et al.*, 'Cyber Terrorism in the process of industry', pp. 17–18.

[166] P. Kerr *et al.*, The Stuxnet Computer Worm: Harbinger of an Emerging Warfare Capability, Congressional Research Service Report for Congress, R41524, p. 3.

The other form of cyberterrorism attack, that involves non-physical destruction, focuses on the modification, manipulation, and destruction of data, websites and information. The result of this form of attack is the negative impact on IT systems by applying the above-mentioned methods. This type of cyber attack affects the integrity, availability, and confidentiality of data. The attack may affect the integrity of data by alteration of the data. The unauthorised access will cause loss of data integrity. If a critical system is attacked and prevents the user from using the system, then the system loses its functionality and effectiveness and loses its availability. This is referred to as loss of the confidentiality of data, which relates to unauthorised disclosure. The unauthorised disclosure of confidential information and data may jeopardise national security, and the impact of it is loss of the confidentiality of data.[167]

A. Intent of a cyber attacker

In fact, defining and proving the intent of a cyber attacker is difficult; particularly for the offences involving the internet, such as cyberterrorism, due to the difficulties in identifying the origin of the attack, the identity of the perpetrator, and the nature of their intent. These measures may remain unknown for a long term during the investigation and prosecution process. A cyber attack has more advantages than a conventional form of attack. Its many advantages are what attract terrorists to use this form of attack. Due to the difficulty in penetrating a program to deploy a cyber attack, it may be said that it only occurs in a case where the attacker intends for it to happen. However, not every act is a terrorist act just because a person intends to break into a computer network; it must also include the intention to influence a government's course of action.

The required elements of a cyberterrorism attack are special intent and special harm which must be fulfilled in order to be considered a cyberterrorism attack and proceed to the sentencing level. This is the element which is determined according to the impact of cyber attack.

---

[167] G. Stoneburner *et al.*, Department of Commerce, National Institute of Standards and Technology, *Risk Management Guide for Information Technology Systems,* NIST Special Publication 800-30, Washington, D.C., 2001. Available at: http://csrc.nist.gov/publications/nistpubs/800-30/sp800-30.pdf (12 Apr 2011).

It would be difficult to establish the necessary level of harm to satisfy the offence elements in cases of cyber terrorist attacks which involve attack against computer systems on the internet, since the types of these attacks are various. Therefore, it is better for countries to gradate damage caused to tangible and intangible property by means of general sentencing rules instead of creating new offences.

There are several differences in the characteristics of cyberterrorism compared to other forms of terrorism. The aim of a terrorist attack is to exhibit the elements of physical harm and cause fear. Thus, the cause of harm can be attained through physical harm and it does not need to go through the trouble of a terrorist attack. Cyberterrorism hurts computerised infrastructure; therefore, the level of a society's vulnerabilities is dependent on their level of dependency on technology and computer networks.

As Richard Clarke stated in 1999, "if you are connected you are vulnerable". Consequently, countries with a high level of technology that result in better defence systems are at risk of being exposed to cyberterrorism. Certainly, the US is more exposed to cyberterrorism than Africa. Although terrorism targets a certain thing that has great potential to cause damage in terms of human life, cyberterrorism hurts a very specific group of people. Another distinction is that cyberterrorism will save on the costs and obstacles of a terrorism attack which requires an executor with all the proper equipment and anti-security verifications. This is because, in a cyber attack, there is no need to purchase a weapon or to be present at the attack's location. All that is needed is a computer and sufficient hacking skills.[168]

i. *The liability of cyber terrorists under the convention on cyber crime*

Generally, cyber crimes are punishable via the Convention on Cybercrime by the Council of Europe. The Convention "intended to create consistency in criminal matters related to internet activities". However, it only criminalises specific activities in order to avoid unwanted effects of cyber crime.[169] Chapter II of the Convention

---

[168] A. Cohen, 'Cyber terrorism: Are we legally ready?' p. 9.

[169] M. G. H. Todd, 'Armed attack in cyber space: Deterring asymmetric warfare with asymmetric definition', *The Air Force Law Review* (2009) 64(65), p. 67.

defines nine different offences in cyber crime which are then divided into four categories, and the offences must be committed "intentionally" to be subject to criminal liability. However, the intention that exists in cyberterrorism is different from that of cyber crime.

An "intention" itself for a cyber attack cannot render an attack a "cyber attack". In other words, the elements that create a cyber attack are not only a mere intention, but also require an intention (*mens rea*) used with the cyber attack to impact on the policy maker.[170] Thus, the policy maker of the member state must consider this element in enacting the law governing cyberterrorism. Thus, applying the term "intention" in the Convention on Cybercrime to address cyberterrorism is vague. It is assumed that cyberterrorism requires "intention" which is covered by the Convention on Cybercrime; therefore, most of the offences in the Convention on Cybercrime are applicable to cyberterrorism. For instance, cyberterrorism attacks may be carried out through illegal access to a computer system without right, through the interception of non-public electronic data transfer, by inflicting damage on the integrity and proper functioning or the use of stored computer data or computer programs.[171] All of these are provided in the Convention on Cybercrime.

The most prevalent method of terrorist attack that is illegal access, is covered by Article 2 of The Convention on Cybercrime. It covers illegal access when there has been access to any part of a computer or the whole computer system without an authorised right. However, it must be committed by "infringing security measures" with the intent of obtaining computer data or any other dishonest intent. It covers all kinds of technical intrusions and terrorists' methods of launching terrorist attacks, especially hacking offences. From a legal point of view, illegal access to computer systems is considered a "primitive offence" against the integrity and confidentiality of computer data and systems.[172] However, this method has always been applied by the terrorists in their attack.

---

[170] A. Cohen, 'Cyber terrorism: Are we legally ready?', p. 33.
[171] Articles 2, 3, 4, 5. Convention on Cybercrime (COC).
[172] P. Brunst, Legal aspect of cyber terrorism, p. 68.

The offences defined in Chapter II of the Convention on Cybercrime are not limited to physical offences. The Convention on Cybercrime is designed to respond to specific manifestations of cyber crime; therefore, it addresses offences that relate to each type of cyber crime individually. Mostly, it applies to cyberterrorism, since cyberterrorism existed when these offences were enacted.[173] The scope of the Convention on Cybercrime goes beyond mere cyber crime, and includes cyberterrorism and all the crimes which involve the computer as well.[174]

Additionally, each member state shall adopt necessary measures to establish illegal access as a criminal offence under its domestic law. Parties agreed on adopting legislation and essential measures in their national law. This is the intent of the Cybercrime Convention, to set up a basic framework so that each party creates their own state criminal legislation, since the Convention has defined those activities broadly. It asked states to criminalise cyber space crimes such as: unlawful access, unlawful interception, interfering with data or systems, and computer fraud or forgery.[175]

It must be accepted that, although the Convention on Cybercrime does not define key terms such as "obtaining" or "altering" clearly, it does indirectly point to the criminal legislation of member states, such as that of the US. This is because, the US "is arguably the largest user of cyberspace; the United States is a leading prosecutor of cyberspace crimes; and many nations work with the United States to solve cyberspace crimes".[176] Countries can use the Convention's provisions as guidelines to develop their domestic legislation and apply the principles and standards in their domestic legal systems.[177]

---

[173] A. Cohen, 'Cyber terrorism: Are we legally ready?', p. 33.

[174] S. L. Marler, 'The convention on cyber-crime: Should the united states ratify?', *New England Law Review* (2002) 37, pp. 183–185.

[175] M. G. H. Todd, 'Armed attack in cyber space: Deterring asymmetric warfare with asymmetric definition', pp. 67–69.

[176] M. G. H. Todd, 'Armed attack in cyber space: Deterring asymmetric warfare with asymmetric definition', pp. 67–69.

[177] S. Schjolberg, A Cyberspace Treaty — A United Nations Convention Or Protocol On Cyber Security and Cyber-crime, *Twelfth United Nations Congress on Crime Prevention and Criminal Justice* Salvador, Brazil, 2010, p. 4.

The fact must be accepted that countries should implement the guidelines in their "substantive criminal law" but they can accept it with certain exceptions according to their situations. The Convention on Cybercrime does not provide precise guidelines for states and it is left to the discretion of the member states. For instance, Article 4(2) of Convention on Cybercrime confers the reservation right to parties concerning the offences in paragraph 1 which "may require that the conduct result in serious harm. The criterion of what constitutes serious harm is left to the discretion of domestic legislations of state parties". However, this may lead to fragmented provisions that do not have the ability to fulfil the purposes of the Convention on Cybercrime in promoting harmonised legal frameworks through an international convention.[178]

1.4 THE DIFFERENCES BETWEEN CYBERTERRORISM AND OTHER RELATED CRIMES

Although there is a major difference between cyberterrorism and related crimes such as "information warfare" and "cyber crime", what usually happens is that the terms overlap. To pave the way towards defining the term "cyberterrorism", there must first be a differentiation between the various meanings of the term to make distinct the difference between "ancillary cyber activities" and "cyber attacks", which uses the internet as the object of attack. The ancillary cyber activities will be considered on Chapter IV.

1.4.1 Cyber Crime and Cyberterrorism

Cyber crime and cyberterrorism are not the same phenomenon, although they refer to unlawful activities in cyber space. Cyber crime can be very broad in scope, and may sometimes involve more factors than just computer hacking. Cyberterrorism is often equated with the use of malicious code. Moreover, a cyberterrorism event may also sometimes depend on the presence of other factors beyond just a

[178] A. Cohen, 'Cyber terrorism: Are we legally ready?', p. 32.

cyber attack.[179] However, there are a number of differences between these two terms.

The significant difference refers to the motive behind a cyber attack. In a case of cyber crime, an unlawful or criminal act occurs where computer technology is either a tool or a target, or both. It is a quite new field of criminological inquiry that comes from the field of criminal justice and it encompasses computer crime or computer-related crime and internet crime.[180]

A cyber criminal is a criminal who uses computers, with or without the internet, to communicate, raise money, recruit new members willing to break the law, and commit other crimes. A cyber crime offence involves "money laundering, fraud and forgery, child pornography, copyright infringements, and security breaches such as hacking, illegal data interception, and system interferences that compromise network integrity and availability".[181] The intention of a cyber criminal is neither political nor social; he/she commits his/her crime through the use of IT.[182] The Council of Europe and the US Department of Justice use the term "cyber crime" for a wide range of crimes involving computers and networks.[183] On the other hand, the definitions for "cyberterrorism" range from being very narrow (using the internet to attack other systems on the internet, resulting in violence against persons or property) to being very broad (also including any other forms of internet usage by terrorists or even conventional attacks that are aimed at IT infrastructure). Generally,

---

[179] C. Wilson, 'Botnets, Computer Attacks, and Cyber Terrorists: Vulnerabilities and Policy Issues for Congress, CRS Report for Congress, United States of America, 2003, pp. 34–37.

[180] J. A. Matusitza, 2006, 'Cyber Terrorism: A Postmodern View of Networks of Terror and How Computer Security Experts and Law Enforcement Officials Fight Them', Ph.D Thesis, University of Oklahoma.

[181] K. Archick, "Cybercrime and Cyber Terrorism". In: J. V. Blane (Ed.), The Council of Europe Convention, pp. 1–6.

[182] Murat Dogrul *et al.*, 'Developing an international cooperation on cyber defense and deterrence against cyber terrorism', 2011, *3rd International Conference on Cyber Conflict*, Cyber Defense, Estonia, 2011, p. 97.

[183] Murat Dogrul *et al.*, 'Developing an international cooperation on cyber defense and deterrence against cyber terrorism', p. 97.

terrorist attacks that are carried out via the internet are especially considered to be "cyberterrorism". Therefore, differences between cyber criminals and terrorist organisations become revealed when the underlying motivation for such attacks is analysed. Cyber criminals often conduct attacks simply to gain monetary income or to demonstrate their "virtual power". This can be achieved, for example, by the following:

(i) Circumventing security measures. Attackers can thereby corrupt the integrity and confidentiality of computer systems and data;
(ii) Rendering systems useless. This can be followed up by further drastic effects if mission-critical IT systems are affected; causing physical harm.

This can be the case if critical infrastructures, such as transportation, power, or water facilities that are connected to an IT system can be manipulated by a perpetrator who has gained access to such a control system. Terrorist organisations, however, typically follow a more long-term perspective. Their general aim is to achieve a primarily political purpose with their actions. Therefore, the following actions are significant to the organisation:

(i) The generation of fear;
(ii) The creation of economic confusion; or
(iii) The discrimination of the political opponent.

Other reasons, however, can also be an underlying agenda for a terrorist act that is being committed via the internet, for example:

(i) The generation of monetary income; or
(ii) The gathering of information on a target (either for a conventional or an electronic attack).[184]

---

[184] P. Brunst, Legal Aspects of Cyber Terrorism, p. 66.

Depending on individual motivation, terrorist aggressions can be performed in different ways. As an example, a hacking attack with the intent to shut down an important system at an airport could be made publicly known in order to arouse fear in the population. However, a hacking attack that is committed in the hopes of gaining information on the automobile route of an important person might be kept secret so as not to endanger future plans for an assassination attempt on that person. In general, it is possible to imagine that all of the general aims mentioned above could also be accomplished with the help of attacks that are committed over the internet.[185]

Along with these terms, there is a phenomenon of cyber crime seen frequently by law enforcement agencies. Cyber crime is a crime committed through the use of IT. It must be pointed out that the physical forms of cyberterrorism, information warfare, and cyber crime often look very much alike.

Imagine that an individual gains access to a hospital's medical database and changes the medication of a pro-business, anti-environmental executive of a Fortune 100 company to one that he or she is dangerously allergic to and also removes the allergy from his or her digital record. The nurse administers the drug and the patient dies. Therefore, which definition should be applied? The answer lies not in the mechanics of the event, but rather in the intent that drove the person's actions. If it was intentionally done, for instance as a result of poor relations between these two people, then it would be murder in addition to a cyber crime. If the executor later would announce that he or she is ready to commit more such acts if their demands would not be met, then it could be labelled as cyberterrorism. If the activities were carried out by an agent of a foreign power, then it could be labelled as information warfare.[186]

Scholars believe that the most important aspect of a cyber attack that has physical consequences is determining the intention of the attacker. The distinction between these terms is extremely important

---

[185] P. Brunst, Legal Aspects of Cyber Terrorism, p. 66.

[186] L. Janczewski *et al.*, *Cyber Warfare and Cyber Terrorism*, Information Science References. In: Kristin Roth (Ed.), 1st Edn, 2008, p. 14.

because there are non-technology-related issues and solutions that will impact any strategy to combat cyber warfare and cyberterrorism.[187]

1.4.2 Cyber Hooliganism and Cyberterrorism

Cyber hooliganism is defined as a criminal action against a computer system that leads to DoS, system destruction, website defacement, stealing user's private mail, etc. Cyber hooliganism can be part of a cyber terrorist's actions, and can also be the work of so-called "script kiddies" (persons who take ready-to-use hacking software created by someone else and run it against a computer system). Cyber hooliganism is essentially non-violent but can cause financial losses. For example, the creation of the 'I love you' virus or destroying the NASA web page are cyber hooliganism acts.

1.4.3 Hacktivism and Cyberterrorism

Scholars have highlighted the importance of hacktivism in the study of cyberterrorism. Hacktivism is electronic civil disobedience or internet activism. Hacktivism consists of writing codes in order to promote political ideology.[188] Hacktivists are cyber protesters; they have political motives and believe that proper use of code will have powerful effects. Hacktivism is the convergence of hacking with activism (used by groups, such as terrorist groups, who use the net to promote their agenda, which can come from any geographical region on the internet, and which can attempt to influence foreign policy anywhere in the world), where the term "hacking tool" is used here to refer to operations that exploit computers in ways that are unusual and often illegal, typically with the help of special software.[189]

---

[187] L. Janczewski *et al.*, *Cyber Warfare and Cyber Terrorism*, p. 15.

[188] M. Milone, 'Hacktivism: Securing the national infrastructure', *Knowledge, Technology, & Policy* (2006) 16(1), pp. 75–103.

[189] D. Denning, 'Activism, hacktivism and cyber terrorism: The internet as a tool for influencing foreign policy', *Georgetown Law University Journal* (2001), p. 15. Available at: http://www.cs.georgetow n/edu/~denning (8 Apr 2010).

Hacktivism includes electronic civil disobedience, which brings methods of civil disobedience to cyber space. However, these groups are not cyber terrorists because they do not cause harm to information systems, websites, and other computer-related materials. In other words, hacktivists do not engage in defacing websites, launching computer viruses, sending worms, or using malicious computer tools. If they do — and since they have political motives — then they become cyber terrorists.

1.4.4 Computer-Assisted Crime and Cyberterrorism

Cyberterrorism is not the same as computer-assisted terrorism. Terrorists can use computer technology, with or without the internet, to support conventional forms of terrorism such as suicide bombings (which are used by al-Qaeda members). They can use websites to communicate with and receive orders from their commanders, obtain important information, carry out missions, propagate their messages, or recruit supporters. In a similar situation, the massive-scale terrorist attacks by the use of email as a medium to communicate information, does not constitute cyberterrorism.

For instance, some experts on terrorism believe that in the 11 September 2001 event, each of the four groups of hijackers did not know each of the other groups, but had communicated with a central "commander" via the internet. This commander might have been a sort of "go-between" or "gatekeeper" between those four groups and might have used email (or a website) to exchange information, procure and channel funding, and organise and order the launching of the attacks against the Twin Towers in New York City and the Pentagon.

Computer-assisted terrorism does not constitute cyberterrorism. "Computer-assisted", like "techno-terrorism", refers to the abundant use of computer technology by terrorists as it adds to their conventional operations. Cyberterrorism, on the other hand, pertains to attacks on information systems, on a nation's computer systems, computer-generated infrastructures, etc.

1.4.5 Information Warfare and Cyberterrorism

As with the terms "terrorism", "cyberterrorism", and "cyber space", there is no widely accepted definition of "cyber warfare". Thus, the definition of information warfare involves different forms. The definition of information warfare encompasses the term "cyber warfare" as well. One definition of information warfare is the definition stated by the US Department of Defense in 1999: "Information operations conducted during time of crisis or conflict to achieve or promote specific objectives over a specific adversary or adversaries". This definition focuses on the military side of information warfare. Another military perspective is that information warfare involves "any action to deny, exploit, corrupt, or destroy the enemy's information and its functions, protecting ourselves against those actions, and exploiting our own military information functions". Although, "information warfare" includes the term "cyber warfare", the interests of cyber warfare are limited to cyber space.

On the other hand, some authors define "information warfare" in a more general way, which is applicable to wider areas, for instance, "Actions taken to achieve information superiority by affecting adversary information, information based processes, and information systems, while defending one's own information, information based processes and information systems". The definition of information warfare from the aspect of computer technology is "the use (and abuse) of computers and high technology appliances to undermine the computing resources of an adversary".[190]

The 2001 Report of the US Congress notes, "Cyber warfare can be used to describe various aspects of defending and attacking information and computer networks in cyberspace, as well as denying an adversary's ability to do the same". A 2006 CRS Report of Congress defined the phrase "computer network attack" as "operations to

---

[190] S. Nitzberg, Conflict and the Computer: Information Warfare and Related Ethical Issues, Telos Information Protection Solutions, *The Proceedings for ETHICOMP*, United Kingdom, 2002, p. 55.

disrupt or destroy information resident in computers and computer networks".

Kevin Coleman, a Senior Fellow and Strategic Management Consultant at the Technolytics Institute, an independent executive think tank, defined "cyber war" as "a conflict that uses hostile, illegal transactions or attacks on computers and networks in an effort to disrupt communications and other pieces of infrastructure as a mechanism to inflict economic harm or upset defenses".

The various definitions of cyber warfare illustrate that military operations in cyber space can be viewed as warfare; the phrase "cyber warfare operations" can be used in analysing the wide range of military operations in cyber space and suggests the following definition: "the use of network-based capabilities of one state to disrupt, deny, degrade, manipulate, or destroy information resident in computers and computer networks, or the computers and networks themselves, of another state". The practical difference between cyberterrorism and cyber warfare is that, cyberterrorism is about causing fear and harm to anyone in the vicinity, while information warfare has a defined target in a war (ideological or declared).

Nowadays, warfare occurs by various new means. These new means can be expressed to be in two forms: kinetic attacks and non-kinetic attacks. A force-on-force engagement would best represent a traditional kinetic attack. The non-kinetic type of attack is not aimed at physical destruction but is designed to impact the adversary's will to fight and decision-making process. Conventionally, this form of warfare is the propaganda or disinformation campaign. Cyber warfare is now a primary tool in the information warfare arsenal to achieve non-kinetic attacks. Cyber warfare has the potential to act as a force multiplier. The cyber terrorist can also take full advantage of this concept. Consequently, a second definition is offered regarding the terrorist's use of cyber space in support of terrorist operations. As a result, both terms, "information warfare" and "cyber warfare", have a specific target in a war, while the fear or harm caused by cyberterrorism is distributed to anyone in the targeted vicinity.

1.4.6 Separation of Ancillary Cyber Activities from Cyberterrorism in the Definition Perspective

Despite cyberterrorism being often interpreted as unlawful actions conducted via computer networks that may cause violence against or generate fear among people, or lead to serious destruction for political or social purposes, this interpretation of cyberterrorism is not clear enough. This is because, when terrorists exercise undirected attacks and use computer and networks as part of their preparations for physical attacks, the question is raised as to whether these actions are considered to be cyberterrorism or not. As we know, for the time being these actions are considered as ancillary cyber activities. This means that although countries include them in their terrorism acts, they do not include them as cyberterrorism crimes. They are just labelled as supportive cyberterrorism activities. One of the main problems in advancing knowledge of cyberterrorism is that many governments are not willing to share their knowledge, since they want to preserve their advanced cyber fighting applications in their own domains.

1.5 CONCLUSION

Scholars offer various definitions of cyberterrorism with different foci, although they exhibit several common elements. Despite these divergent definitions, views, and foci, the most accepted definition can only be established by integrating all elements. As previously mentioned, cyberterrorism notions are defined based on two categories: intent-based definition and effect-based definition. The intent-based definition focuses on the political intention of the attacker. The *mens rea* of the attacker plays a vital role in constructing the cyberterrorism notion. The effect-based definition focuses on the result of the attack, that is, the attack must result in serious harm to persons or property. The most complete definition combines the two elements but does not overlap with related cyber crime.

CHAPTER II

THE CHALLENGES FACED BY INTERNATIONAL ORGANISATIONS IN CURBING CYBERTERRORISM

2.1 INTRODUCTION

Governments of various countries, after hard attempts to fight cyberterrorism alone, have reached a point of understanding that bilateral and multinational cooperation is the most suitable method to counter any transnational crime such as cyberterrorism. Cyberterrorism is a transnational crime and international threat and national regulations alone cannot prevent it. The need for an international organisation to prevent and defend from cyberterrorism attacks is urgent.

Since cyberterrorism and all crimes related to the internet are inherently transnational, nations should work together to deter these related crimes. Also, legal enforcement by states is not worthwhile in a global world because of the limited jurisdiction that law enforcement has.

International cooperation to confront cyberterrorism has different forms among government and law enforcement agencies. The cooperative efforts are divided into three types: international and global effort, multilateral and multinational effort, and regional effort. Due to its transnational nature the efforts to confront this offence must be international. Because countries have different laws and rules of extradition and laws governing computer crime, the need for multilateral cooperation emerges. Scholars assert that effective preventive methods to respond to cyber attacks and cyberterrorism must be global; therefore international cooperation amongst relevant organisations is the best one. Bilateral cooperation has fewer advantages in contrast with multilateral cooperation. Firstly, the scope of this cooperation is narrower in the number of countries and secondly these treaties do not cover cyber crime specifically and furthermore application of these treaties is time-consuming since they involve much paperwork.[1]

2.1.1 Objective of the Chapter

This chapter aims to study the available international mechanisms for the purpose of establishing cooperation in responding to cyberterrorism. The current international cooperation and responses for cyberterrorism are reviewed and analysed. The responses of international organisations to cyberterrorism cases are also considered.

2.2 EFFORT TAKEN BY INTERNATIONAL ORGANISATIONS

Cyberterrorism applies to offences across borders; therefore the proper response which has the potential to combat cyberterrorism should be transnational. Countries must bind together with multilateral and bilateral cooperation. In this section, the most effective organisational structures will be considered.

2.2.1 The United Nations (UN)

The UN is an international organisation founded in 1945 after the Second World War that is committed to keeping international peace and security. The work of the UN in its various methods reaches every

---

[1] L. Xingan, 'International actions against cybercrime: Networking legal systems in the networked crime scene', *Webology* (2007) 4(3).

corner of the world. The UN is the lead organisation that tries to coordinate and seek cooperation in dealing with the problem of international terrorism.[2] In addition, the UN, as a global organisation that forms a forum of 191 member states, does not confine itself to a certain area, and unlike other organisations, does not limit itself to certain states. The main goal of the UN is to keep maintaining international peace and security. Nevertheless, it establishes many specialised agencies and programmes.[3]

According to the UN Charter and related organisations, they work on a broad range of fundamental issues "from sustainable development, environment and refugees protection, disaster relief, counter-terrorism, disarmament and non-proliferation, to promoting democracy, human rights, gender equality and the advancement of women, governance, economic and social development and international health, clearing landmines, expanding food production, and more"[4] Similarly, the UN has enacted several conventions on terrorism that include: The Offences Committed On Board Aircraft, The Unlawful Seizure of Aircraft, Crimes against the Safety of Civil Aviation, Crimes against Internationally Protected Persons, Taking of Hostages, Unlawful Use of Nuclear Material, Unlawful Acts against the Safety of Maritime Navigation, Unlawful Acts against the Safety of Fixed Platforms Located on the Continental Shelf, Making of Plastic Explosives for the Purpose of Detonation, Terrorist Bombings, Financing of Terrorism and Nuclear Terrorism.[5]

Furthermore, within Resolutions 55/63 (2000) and 56/121 (2001) on Combating the Criminal Misuse of Information Technology (IT) the value of the Group of Eight (G8) Principles were noted. Its member states are urged to consider these principles; further, there are

---

[2] D. Yaman, The United Nations and Terrorism, Legal aspects of terrorism, In *Legal Aspect of Combating Terrorism*, Centre of Excellence Defence against Terrorism. IOS Press Publications, Turkey, p. 11.

[3] D. Yaman, The United Nations and Terrorism, p. 11.

[4] United Nations (1945) (UN) Available at: http://www.un.org/en/aboutun/index.shtml (22 Sept 2011).

[5] United Nations (1945) (UN) Available at: http://www.un.org/en/aboutun/index.shtml (5 May 2010).

also other Resolutions[6] calling on member states "to promote the multi-lateral consideration of existing and potential threats in the field of information security, as well as possible measures to limit the threats". These resolutions all have the same motive which is to improve cyber security awareness at both the international and the national levels.[7]

Following the tragedy of 11 September 2001, the Security Council Resolution 1373 moved forward to fight against terrorism. The aim of this Resolution was to counter-terrorism efforts and it was adopted on 28 September 2001. It declared international peace and security and imposed an obligation on all UN member states to support the prevention of terrorism as well as criminalising terrorist activity, terrorist financing, and supporting terrorist activity. It obliged all member states to cooperate with other government and international organisations to deny safe havens for terrorists. It directed efforts in different aspects of fighting against terrorism.[8] The UN Security Council Resolution 1373 (2001) imposes mandatory obligations to block terrorist funds, to prevent terrorist activity, to cooperate on judicial and extradition questions, and to coordinate on the related issues of transnational crime, illicit drugs and arms trafficking.[9] The UN International Convention for The Suppression of The Financing of Terrorism, which was adopted by the General Assembly of the UN in Resolution 54/109 in 1999, also promotes international cooperation. Requests for cooperation may not be refused on political grounds, according to Article 14 of this Convention.[10]

In addition, it established a committee to monitor the implementation of these measures and it criminalised any kind of active and

[6] Developments in telecommunications and information in the context of international security, 53/70 (1998), 54/79 (1999), 55/28 (2000), 56/19 (2001), 57/53 (2002), 57/239 (2002), 58/32 (2003), and 58/199 (2003).

[7] L. Xingan, 'International actions against cybercrime: Networking legal systems in the networked crime scene'.

[8] About The Anti-Defamation League. Available at: http://www.adl.org/Terror/tu/tu_38_04_09.asp (22 Sept 2011).

[9] D. Yaman, The United Nations and Terrorism, p. 11.

[10] The UN International Convention for The Suppression of The Financing of Terrorism of 1999, Un treaty series Reg. No. 38349, adopted by the General assembly of the United Nations in resolution 54/109 on 9.12.1999.

passive assistance to terrorism and demanded nations to become part of the relevant international resolutions and conventions relating to terrorism. The UN set up a Counter Terrorism Cell to monitor the implementation of Resolution 1373 and to assist nations in developing the required capabilities.[11]

UN Security Council Resolution 1566 (2004) requires states to fully cooperate in the fight against terrorism, and Resolution 1624 (2005) calls upon states to "prohibit by law incitement to commit a terrorist act or acts".[12] Accordingly, these two Resolutions will be discussed. UN Security Council Resolution 1566 was constructed based on Chapter VII of the UN Charter which considers the act of terrorism seriously and condemns terrorism in all of its forms. It calls on the states to cooperate fully with the Counter Terrorism Committee (CTC) which was established pursuant to Resolution 1373 (2001). This Resolution provides an internationally recognised definition of "terror" for the first time which seems to provide an inclusive ban on all forms of violence that intentionally target civilians, regardless of the motive, as well as calls on countries to prosecute terrorists. The UN set up the CTC to monitor the implementation of Resolution 1373 and to assist nations in developing the required capabilities.[13] The CTC is known as the "centre of global efforts to fight terrorism". It has imposed sweeping legal obligations on UN member states to combat global terrorist threats. It goes beyond the other existing counter-terrorism treaties (that bind those who have voluntarily become parties to them by creating uniform global obligations) by requiring every country to freeze the financial assets of terrorists and their supporters.[14]

---

[11] D. Yaman, The United Nations and Terrorism, p. 11.

[12] R. F. Perl, Terrorist Use of the Internet: Threat, Issues, and Options for International Cooperation, *Organization for Security and Cooperation in Europe*, 2008, p. 5.

[13] R. F. Perl, Terrorist Use of the Internet: Threat, Issues, and Options for International Cooperation, p. 5.

[14] E. Rosand, The un-led multilateral institutional response to jihadist terrorism: Is a global counterterrorism body needed? *Journal of Conflict and Security Review* (2006) 3, p. 405.

The task of the CTC is to monitor the member-states' implementation of Security Council Resolution 1373 where necessary as well as to provide these states with technical assistance in preventing cyber and conventional terrorism. Security Council Resolution 1373 was adopted by establishing the Counter-Terrorism Executive Directorate (CTED) to strengthen the CTC's efforts. The CTED assists states in meeting the 1373 resolution's obligations. Although both of these UN bodies face many challenges, they have exhibited impressive progress.

The CTC followed a special approach to implement resolution 1373, but this approach may hinder its monitoring duties. This approach includes the examination of a state's anti-terrorism legislation and program as well as the monitoring of the implementation of the resolution. However, the committee has failed to challenge countries that support terrorists. For instance, the committee failed to ask Syria about its support of terrorist organisations such as Hamas, Hezbollah, and the Islami Jihad.[15] The CTC also has no enforcement power, and the applicability of its ruling depends on cooperation and transparency. Resolution 1617 (2005) calls on states to combat terrorism in all its forms in accordance with the Charter of the UN and also stresses that states must ensure that any measures taken to combat terrorism comply with all their obligations under international law, and should adopt such measures in accordance with international law, in particular international human rights law, refugee law, and humanitarian law.[16]

In 2008, the UN General Assembly adopted Resolution A/RES/2321 on cyberterrorism, focusing on enhancing public awareness and calling for a standard punishment for these types of attacks. In addition, in 2010, the General Assembly adopted a resolution on the "creation of a global culture of cyber security and taking stock of national efforts to protect critical information infrastructures",

[15] A. Bianchi, 'Assessing the effectiveness of the UN Security Council's anti-terrorism measures: The quest for legitimacy and cohesion', *European Journal of International Law* (2006) 17(5). Available at: http://ejil.oxfordjournals.org/content/17/5/881.full (29 May 2013).

[16] International Law and Cyber Defence, United States and Infrastructure, p. 3.

encouraging its member states to share best practices and measures in cyber security. The potency of a Security Council Resolution, which is legally binding and vested by Article 25 of the UN Charter, cannot be under-emphasised. It is a strong and effective implementation tool which has global application to all UN members.

For instance, Resolution 1373 borrows various obligations from existing counter-terrorism conventions and applies them to all UN member states, without the need for them having to sign those conventions. What is more, a Security Council Resolution can be established in a short period of time, compared with the period of time it takes for a treaty to be adopted. Drafting a treaty between states might take many years. Considering these advantages, although a Security Council Resolution cannot create a complete international counter-terrorism instrument, it can still be a very effective tool.[17]

Following the September 11 attacks, the UN developed mechanisms to combat conventional and cyberterrorism despite the lack of an internationally accepted definition for cyberterrorism. The UN has had a valuable role in combating conventional and cyberterrorism. The coalition against cyberterrorism is coming together owing to regional efforts. According to the Vienna Convention on the Law of Treaties (known as the "Treaty on Treaties"), parties can choose to interpret sources other than the text of a treaty as long as they agree that those sources provide interpretive information that is authoritative.[18] Therefore, the text of a treaty may be expanded, either explicitly or implicitly, to cope with rapid technological changes so as to cover new circumstances and thus, legislation may be amended to reflect existing legal circumstances. In this way, most of the international counter-terrorism conventions can be applied to cyberterrorism. Although the UN has published charters on war and terrorism, establishing laws and conventions pertaining to cyberterrorism is a

[17] J. Trahan, 'Terrorism convention: Existing gaps and different approached', *New England International and Comparative Law Annual* (2002) 8, p. 221.

[18] A. Cohen, 'Cyber terrorism: Are we legally ready?', *Journal Of International Business and Law Review* (2010) 9(1), p. 86.

complicated process, and the UN has not created a comprehensive convention that covers all acts of terrorism as yet.[19]

Furthermore, the victims of terrorism and cyberterrorism likely fall within the meaning of Crimes against Humanity (CAH). However, international communities such as the UN have failed to adopt a specialised convention on CAH. Although, there were some tribunals after World War II which contained "crimes against humanity", international communities have not established a special convention for such kinds of crime.[20]

However, there are 17 specific conventions (including its complimentary) and major legal instruments which deal with terrorist activities and which may be applicable to cyberterrorism. These are:

(1) The 1963 Convention on Offences and Certain Other Acts Committed on Board Aircraft;
(2) The 1970 Convention for the Suppression of Unlawful Seizure of Aircraft;
(3) The 2010 Protocol Supplementary to the Convention for the Suppression of Unlawful Seizure of Aircraft;
(4) The 1971 Convention for the Suppression of Unlawful Acts against the Safety of Civil Aviation;
(5) The 1973 Convention on the Prevention and Punishment of Crimes against Internationally Protected Persons;
(6) The 1979 International Convention against the Taking of Hostages;
(7) The 1980 Convention on the Physical Protection of Nuclear Material;
(8) The 1988 Protocol for the Suppression of Unlawful Acts of Violence at Airports Serving International Civil Aviation;

[19] O. Yen Nee, International Responses to Terrorism: The Limits and Possibilities of Legal Control of Terrorism by Regional Arrangement with Particular Reference to Asean, Institute of Defence and Strategic Studies, Singapore, 9–10, 2002, p. 5.
[20] M. Bassiouni, 'Crimes against humanity: The case for a specialized convention', *Washington University Global Studies Law Review* (2010) 9(4), p. 575.

(9) The 1989 Supplementary to the Convention for the Suppression of Unlawful Acts against the Safety of Civil Aviation;
(10) The 1988 Convention for the Suppression of Unlawful Acts against the Safety of Maritime Navigation;
(11) The 2005 Protocol to the Convention for the Suppression of Unlawful Acts against the Safety of Maritime Navigation;
(12) The 1988 Protocol for the Suppression of Unlawful Acts against the Safety of Fixed Platforms Located on the Continental Shelf;
(13) The 1991 Convention on the Marking of Plastic Explosives for the Purpose of Detection;
(14) The 1997 International Convention for the Suppression of Terrorist Bombings;
(15) The 1999 International Convention for the Suppression of the Financing of Terrorism;
(16) The 2005 International Convention for the Suppression of Acts of Nuclear Terrorism; and
(17) The 2010 Convention on the Suppression of Unlawful Acts Relating to International Civil Aviation.[21]

For the time being, the only international UN body that specialises in dealing with cyber attacks is the International Telecommunications Union (ITU), which has been attempting to set up a focus group to establish a minimum reference point against which network operators can access their security. The Secretary General of the ITU, Dr. Hamadoun Touré, has proposed an international cyber peace treaty, and continues to advocate this idea despite resistance to it. In addition, Dr. Touré has proposed a "common code of conduct against cybercrime" which would obligate countries to: (i) protect their citizens against cyber criminals, (ii) deny safe haven to terrorists or criminals within their territories, and (iii) not to attack another country first.

[21] UN Action to Counter Terrorism, International Legal Instruments to Counter Terrorism. Available at: http://www.un.org/terrorism/instruments.shtml (23 Aug 2012).

As will be discussed in the following section, at this time, the only international treaty existing to deal with cyber crime is the Council of Europe's Cybercrime Convention, which has been signed by 43 countries (the majority of which are technologically-advanced). Meanwhile, it is encouraging to note that many other countries have initiated efforts to adopt the principles of this Convention into their own legal frameworks.

In the same vein, President Bill Clinton created the Commission of Critical Infrastructure Protection in 1998, as a grouping of electricity, communications, and computer bodies, whose aim is to defend critical infrastructure from physical and cyber attacks. Subsequently, President George W. Bush established the National Security Agency with a large cybernetics strike force. In August 2011, the UN launched the African Center for Cyber Law and Cybercrime Prevention in order to monitor cyber space and cyber crime in African jurisdictions. By the same token, in 2011 in Asia, Singapore announced the first cyber security training and accreditation program with the view to teaching cyber security professionals how to defend countries from hackers.[22]

2.2.2 The Organization for Security and Cooperation in Europe (OSCE)

The OSCE is one of the branches of the UN which hosts numerous activities in counter-terrorism. The OSCE is initiated under Chapter VIII of the UN Charter. The legal framework is found under anti-terrorism conventions and protocols and relevant UN Security Council Resolutions, particularly Resolution 1373 (2001) which calls upon all states "to co-operate, particularly through bilateral and multilateral arrangements and agreements, to prevent and suppress terrorist attacks and take action against perpetrators of such acts".[23] It establishes a body that is called the Action against Terrorism Unit (ATU) and it is a focal point and facilitator of OSCE counter-terrorism activities.

---

[22] M. Conway, 'Terrorist use of the internet and fighting back', *Security and Information International Journal* (2006) 9, p. 25.

[23] Action against Terrorism Unit, Organization for Security and Cooperation in Europe. Available at: www.OSCE.org (4 June 2013).

The ATU takes measures to strengthen the international legal framework against terrorism incorporated within the UN Office on Drug and Crime (UNODC). The ATU assists participating states to draft legislation in criminalising the terrorist offences in their domestic legal systems. This is because they have discovered that counter-terrorism activity cannot be effective unless they have a strong legal framework. It attempts to enhance international cooperation in criminal matters of terrorism. Since 2005, by joining the UNODC, many participating states have discovered that, due to the fact that terrorism-related crime is transnational and cross-border, legal cooperation is required to investigate and prosecute perpetrators.[24]

Another important role of the ATU lies in combating terrorists' use of the internet by identifying potential counter-measures. The ATU also hosts a wide range of OSCE events on terrorists' use of the internet for scholars and experts to work on. The OSCE follows these activities to enhance cyber security in a comprehensive manner. The OSCE is fast becoming a leader in combating the threat of cyber attacks. However, this organisation is confronted with several challenges similar to other international communities. The OSCE suffers from a lack of support from all countries and must enable their members to strengthen their national forensic capabilities for law enforcement agencies' fight against cyberterrorism. Providing resources such as training and equipment to assist countries in dealing with cyberterrorism is also very difficult.

2.2.3 Interpol

Interpol is an international police organisation and is the largest one in the world. It was created in 1923 to facilitate cross-border police cooperation even where diplomatic relations do not exist between particular countries and to support and assist all organisations, authorities, and services whose mission is to prevent or combat international crime. Interpol began its efforts to improve its counter-cyber crime capacity at the international level very early. A 1981 survey of

[24] United Nations Office on Drugs and Crime, United Nations Convention against Transnational Organised Crime and the Protocols Thereto, New York, 2004, pp. 55–60.

members on cyber criminal law recognised dilemmas in the application of existing legislation.[25] Based on the recognition of the legal gaps between countries and gaps between the legal framework and criminal phenomena, Interpol expanded its task to both law enforcement and legal harmonisation.[26]

As discussed above, Interpol works on international crimes. One of the main issues among these crimes is financial and high-tech crimes — such as currency counterfeiting, money laundering, intellectual property crime, payment card fraud, computer virus attacks, and cyberterrorism — which can affect all levels of society.[27]

Interpol can be a resource for countries that suffer a cyber attack. For instance, when Estonia suffered a cyber attack, Interpol was another international avenue for Estonia to investigate this case. This was due to the fact that Interpol has the ability to rely on a state's law enforcement body — if it is a member of Interpol — and request professional assistance. The parties must have a legal basis in an international multilateral statutory act or have a bilateral agreement to carry out a certain procedural act through Interpol. However, it appears that making a request to Interpol is not much different to a request made to NATO (it will be discussed in next section, under regional level of effort), since the functioning of Interpol relies on the same treaty regime to conduct a pre-trial investigation; therefore, the possibility of encountering the same problems is very high.

In addition to this, Interpol has formed a group as a public–private partnership to develop strategies to combat transnational organised intellectual property crime and encourage National Central Bureaus (NCBs) and national law enforcement to enact more intellectual property crime law enforcement to combat transnational organised intellectual property crime. What is more, since intellectual

---

[25] S. Schjolberg *et al.*, Computer-Related Offences, *Conference on the Challenge of Cybercrime*, Council of Europe, Strasbourg, France, 2007. Available at: http://cybercrimelaw.net/documents/strasbou rg.pdf (22 Sept 2011).

[26] L. Xingan, 'International actions against cybercrime: Networking legal systems in the networked crime scene'.

[27] Financial and high-tech crimes. Available at: http://www.interpol.int/Public/FinancialCrime/Default.asp (12 May 2010).

property crime is a high profit, low risk crime, it inevitably motivates criminals, especially terrorist organisations, to utilise counterfeit and pirated funds for their activities through intellectual property crime. Interpol cooperates with law enforcement authorities in the fight against transnational organised intellectual property crime.[28]

The Interpol General Secretariat has harnessed the expertise of its members in the field of Information Technology Crime (ITC) through the vehicle of a "working party" or a group of experts. In this instance, the working party consists of the heads or experienced members of national computer crime units. These working parties have been designed to reflect regional expertise and exist in Europe, Asia, America and in Africa.[29] The main task of the central body is to harmonise the different regional working party initiatives. Apparently, legal harmonisation is one of Interpol's important tasks in working toward an effective law enforcement environment.[30] This task is the only way that might bring countries together to punish cyberterrorism crimes.

Interpol created an anti-terrorism section in September 2002, in the wake of an alarming rise in international terrorist attacks, called the Fusion Task Force (FTF). Its primary objectives are to identify active terrorist groups and their membership, solicit, collect, and share information and intelligence, provide analytical support and enhance the capacity of member countries to address the threats of terrorism and organised crime.[31] Six regional task forces have been created in regions considered to be particularly susceptible to terrorist activity; Project Pacific (Southeast Asia), Project Kalkan (Central Asia), Project Amazon (South America), Project Baobab (Africa), Project Nexus (Europe), and Project Middle East.

---

[28] Intellectual Property Rights Program. Available at: http://www.interpol.int/Public/Financialcrime/intellectual Property (12 May 2010).

[29] Information Technology Crime. Available at: http://www.interpol.int/public/technologyCrime/default.asp (12 May 2010).

[30] Available at: http://www.interpol.net/Public/TechnologyCrime/WorkingParties/Default.asp (23 May 2010).

[31] Fusion Task Force. Available at: http://www.interpol.int/Public/FusionTaskForce/default.asp (12 May 2010).

The FTF has a website for its members and provides all information on its working group meetings, including presentations and analytical reports, photo boards of suspected terrorists, and notices and diffusion lists. In January 2008, 545 users accessed the FTF restricted web site.[32] Interpol has identified public safety and terrorism as a priority crime area and countries can benefit from Interpol's unique position in the international law enforcement community in the fight against terrorism. The Interpol officials involved with the FTF are all terrorism specialists seconded from their home countries.[33]

The first INTERPOL World was held in Singapore from 14 to 16 April 2015. IGC will work as a global cooperation hub between law enforcement agencies. The IGC is being built to achieve Interpol's objective to become a global reference point in research and development. The facility aims to enhance Interpol's capability to address 21st century crime and strengthen international policy. Interpol's supreme governing body decided to establish the IGC in November 2010 and endorsed the proposal during the body's 79th session in Doha, Qatar. This new complex brings the east and west together to serve global security and create a global entity. The Singapore facility will be built as a complement to the general secretariat headquartered in Lyon.[34]

The efforts of international organisations are more appropriate than the efforts of regional organisations. The existing international conventions for the harmonisation of national laws are applicable to the prosecution of cyberterrorism. For the time being, these international organisations are the best methods in preventing and prosecuting cyberterrorism. The major weakness of international instruments is the lack of ratification and implementation. The broad acceptance of countries is important for international instruments to fight against cyberterrorism. If the UN establishes that terrorists who use the internet fall under cyberterrorism, this notion would be better for the fight

[32] Fusion Task Force. Available at: http://www.interpol.int/Public/FusionTaskForce/default.asp (12 May 2010).

[33] Fusion Task Force. Available at: http://www.interpol.int/Public/FusionTaskForce/default.asp (12 May 2010).

[34] INTERPOL, The Interpol Global Complex in Singapore. Available at: https://www.interpol.int/Public/Icpo/IGC/igc.asp (20 Aug 201).

against cyberterrorism. Considering that legal measures in cyberterrorism serve a critical function, we believe that having a common definition for cyberterrorism with unique characteristics will ease the establishment of cooperation between countries. Countries can achieve real cooperation on cyberterrorism by becoming involved in international communities. International communities must provide conditions to promote a general consensus between countries.

2.3 REGIONAL LEVEL EFFORT AND COOPERATION

2.3.1 The European Union

The 1957 Treaty of Rome established the European Economic Community, which in turn evolved into the EU established under the Treaty of Maastricht in 1992. After the terrorist attack on Madrid in 2004, the EU and its members promised to do everything in their power to combat all forms of terrorism. Thereafter, the European Council approved the decision of the European Parliament to declare 11 March as the European Day Commemorating the Victims of Terrorism. On 1 May of 2001 the European Council came up with a Council Directive and invited the European Commission to compensate crime victims and ensure an allocation from its budget for supporting victims of terrorism.[35]

In December of 2004, the European Council called on member states to ratify the Convention on Mutual Assistance in Criminal Matters, its protocol and the three protocols to the Europol Convention. In addition, its framework implemented other aspects such as traffic data by service providers, cross-border pursuit, and the exchange of information on conviction for terrorist offences. The Council adopted necessary measures for Council regulations to identify new and applicable functions for the Schengen Information System (SIS).[36]

---

[35] Declaration on Combating Terrorism, Brussels: Council of the European Union, p. 18. Available at: http://www.consilium.europa.eu/ueDocs/cms_Data/docs/pressData/en/ec/79637.pdf (2004).

[36] Declaration on Combating Terrorism, p. 18. Available at: http://www.consilium.europa.eu/ueDocs/cms_Data/docs/pressData/en/ec/79637.pdf (2004).

The European Council invited the European Commission to bring forward proposals to the European Council meeting in relation to exchange of personal information, for example DNA, fingerprints, and visa data, for the purpose of combating terrorism. The European Commission's proposals should also include provisions to enable national law enforcement agencies to have access to the EU systems. The European Commission is also invited to consider the criteria that should be applied corresponding to Article 96 of the Schengen Convention in relation to certain persons reported for the purposes of being refused entry.[37]

In addition, the EU works on strengthening border control and security to combat terrorism and cyberterrorism as a new kind of terrorism. The EU instructed the Council to adopt the incorporation of biometric features into passports and visas and also an integrated system for the exchange of information on stolen and lost passports having resources to the SIS and Interpol databases (the EU have the Interpol database and resources of the SIS). In addition, the EU has strategic objectives to combat terrorism. The EU strengthens international consensus and combats terrorism by enhancing international efforts as well as enhancing the capability of member states to protect against terrorist attacks. In 2002, a proposal was presented by the European Commission to the European communities for a Council Framework on Decision Making against information systems. The proposal criminalises offences of illegal access, illegal interface, illegal data interface, and instigation, aiding, abetting, or attempting these offences.[38]

The European Police Office (Europol) is part of the EU and was created to increase the efficacy of cooperation among police agencies of EU member states. It consists of two types of officers: European Liaison officers, and Europol staff officers. The function of Europol is somewhat like Interpol. It supports the "operational activities of

---

[37] Article 96 of the Schengen Convention states that the refusal of entry for aliens must be on the basis of a national alert resulting from the decisions taken by the authorities in accordance with rules of procedure laid down by national law.

[38] Common security and defence policy, Fact Sheets on the European Union-2013, http://www.europarl.europa.eu/ftu/pdf/en/FTU_6.1.3.pdf (15 Aug 2013).

national law enforcement officials" and attempts to fight against cyber crime and cyberterrorism. The focus of Europol is on organised crime. Europol representatives provide technical expertise for future investigation within the EU, analysis of criminal intelligence, and reports of criminal activity. Additionally, it also provides an investigative and operational role.[39] The Europol function in preventing cyberterrorism is to support the activities of national law enforcement agencies in fighting against cyber crime and cyberterrorism. In doing so, the agency provides analyses of criminal intelligence and technical expertise for the investigation of cyber crime and cyberterrorism offenses within the EU.

The main role of Europol in preventing cyberterrorism offences is to expand the investigation process and operational role to delineate the offence and perpetrators. However, the agency may confront problems in investigating cyberterrorism cases in the EU zone. Considering that cyberterrorism offenses begin in one place and end in another place, the investigation process confronts many problems in addressing cyberterrorism. The cyber attack that happened in Estonia in 2007 is a good example of this type of problem. Estonia did not have proper laws to prevent and prosecute cyber attacks.

2.3.2 The Council of Europe (CoE)

The CoE has been dedicated, since 1949,[40] to upholding human rights, the rule of law and pluralist democracy. It is an intergovernmental organisation.[41] It is determined to combat terrorism which

---

[39] R. Broadhurst, 'Developments in the global law enforcement of cyber-crime', *An International Journal of Police Strategies and Management* (2006) 29(6), p. 425. Available at: http://dx.doi.org/10.1108/ 13639510610684674 (14 Apr 2012).

[40] The Council of Europe (CoE) and European Union (EU) are two distinct bodies established with the aim of enabling Europe and its member nations to prosper. The two organisations have their own sets of goals and objectives. Each of these bodies has its own subdivisions that specialise in various economic areas or uphold certain democratic concepts to ensure the utmost respect for human rights.

[41] Difference between European Union and Council of Europe. Available at: http://www.differencebetween. .net/business/organizations-business/difference-between-the-european-union-and-council-of-europe/ (3 May 2013).

repudiates these three fundamental values. The CoE has worked in this field since the 1970s but its efforts were stepped up in 2001 following the unprecedented terrorist attacks in the US.[42] The CoE's activities in the fight against terrorism are built on three cornerstones:

(i) Strengthening legal action against terrorism;
(ii) Safeguarding fundamental values; and,
(iii) Addressing the causes of terrorism.[43]

Following the attacks of 11 September 2001 on the US, the Committee of Ministers of the CoE agreed to take steps rapidly to increase the effectiveness of the existing international instruments within the Council on the fight against terrorism by, *inter alia*, setting up a Multidisciplinary Group on International Action against Terrorism (GMT).[44] The aim of this group is identified in a range of priorities and it is at the forefront of action by the CoE against terrorism. The main aims of this group are implementing special investigation techniques, protecting witnesses and collaborators of justice, engendering international cooperation on law enforcement, taking pre-emptive action to prevent terrorists from accessing funding sources and scrutinising identity documents which arise in connection with terrorism.[45]

---

[42] Council of Europe, Human Rights and Legal Affairs. Available at: http://www.coe.int/t/e/legal_co-operation /fight_against_terrorism/1General/ (17 May 2010).
[43] Council of Europe, Human Rights and Legal Affairs. Available at: http://www.coe.int/t/e/legal_affairs/legal_co-operation/fight_against_terrorism/1_General/ (17 May 2010).
[44] GMT is a review work of a governmental committee which was drafted following the Convention on the Prevention of Terrorism in 1977 which brought experts together from the 45 Council of Europe member states and observer states. One of the principal functions of the GMT is to make profiles of each member's legislative and institutional counter-terrorism plans. These profiles are short reports on member states' counter-terrorism measures.
[45] Council of Europe, The Multidisciplinary Group on International Action Against Terrorism. Available at: http://www.coe int/t/dlapil/ codexter/3_codexter/the_gmt/gmt_more_EN.asp (22 Sept 2011).

Since 1949, the CoE has been an intergovernmental organisation dedicated to upholding human rights, the rule of law, and pluralist democracy.[46] The Committee of Experts on Terrorism (CODEXTER) has now replaced GMT as the coordinating committee for the implementation of actions of CoE against terrorism following the 24th and 25th Conferences of European Ministers of Justice held in Moscow, Russia in 2001 and Sofia, Bulgaria in 2003, respectively.

CODEXTER coordinates the implementation of activities in protecting, supporting and compensating victims of terrorist acts, assessing the effectiveness of national judicial systems in their responses to terrorism, supporting the upgrading of member states' legislative and institutional counter-terrorism capacities and technical cooperation and profiling different countries' counter-terrorism capacities.

In 2005, CODEXTER identified the following additional priority areas, namely to identify young persons as terrorist offenders and targets of terrorist propaganda, to counter cyberterrorism and misuse of the internet for terrorist purposes, to uncover insurance schemes to cover terrorism-related damages, the denial of residence to foreign terrorists, and combating and preventing terrorism through culture.[47]

The CoE set its focus area on cyberterrorism and the main aim of CODEXTER is combating cyberterrorism. CODEXTER has been surveying the situation in member states to evaluate whether existing international instruments are sufficient to respond to cyber threats. CODEXTER has pursued its consideration of the use of the internet for terrorist purposes and the notion of cyberterrorism in subsequent meetings, on 10th June 2006, 11th December 2006, and 12th April

---

[46] The Council of Europe (CoE) and European Union (EU) are two distinct bodies established with the aim of enabling Europe and its member nations to prosper. The two organisations have their own sets of goals and objectives. Each of these bodies has its own subdivisions that specialise in various economic areas or uphold certain democratic concepts to ensure the utmost respect for human rights.

[47] Council of Europe, Humans Rights and Legal Affairs. Available at: http://www.coe.int/t/e/legal_affairs/legal_co-operation/fight_against_terrorism/1_General/ (17 May 2010).

2007.[48] CODEXTER has concluded at the end of these meetings that firstly, the use of the internet for terrorist purposes includes several elements:

a. Attacks via the internet that cause damage not only to essential electronic communication systems and IT infrastructure but also to other infrastructures, systems and legal interests, including human life;
b. Dissemination of illegal content including threatening terrorist attacks such as: inciting, advertising, and glorifying terrorism, fundraising for and financing of terrorism, training for terrorism, and recruiting for terrorism; as well as,
c. Other logistical uses of IT systems by terrorists, such as internal communication, information acquisition, and target analysis.

The CoE drew up the Convention on Cybercrime and the Convention on Prevention of Terrorism for fighting cyberterrorism and other terrorist use of the internet. However, serious threats of cyberterrorism are not adequately covered either by the above mentioned conventions (created for conventional terrorism) or other CoE conventions (created for cyberterrorism). This deficit is also not compensated for by other international organisations. Given that the Convention on Cybercrime and the Convention on the Prevention of Terrorism are the most significant international instruments in this case and the number of member states is insufficient, the signatures, ratification, and implementation of these two conventions should be supported.[49]

---

[48] Council of Europe, Humans Rights and Legal Affairs. Available at: http://www.coe.int/t/e/legal_affairs/legal_co-operation/fight_against_terrorism/4_Theme_Files/ (21 May 2010).

[49] The Council of Europe drew up these two conventions to achieve greater unity between member states and to fight against cyber-crime, cyber terrorism, and the use of internet for terrorist purposes.

A. Convention on Cybercrime

This Convention was released in April 2000, but it has been revised after that date several times. The Convention on Cybercrime was adopted in 2001 by the CoE, a consultative assembly of 43 countries, based in Strasbourg. The Convention on Cybercrime, put into effect in July 2004, was the first and only international treaty to deal with breaches of law over the internet or other information networks. The Convention on Cybercrime seeks to combat cyber crime by various approaches such as harmonising national laws, improving the instruments of investigation and investigation ability, and increasing international cooperation and efforts to criminalise terrorism using the internet. In the virtual world a computer works as a tool or target for cyber criminals and cyber terrorists. Cyber terrorists utilise modern high-tech methods to commit their cyber attacks that include botnet attacks, digital piracy, malicious spreading of viruses, attacks on critical information infrastructure by criminal groups and hacking.[50]

The Convention requires participating countries to update and synchronise their criminal laws against hacking, infringements on copyright, computer-facilitated fraud, child pornography, and other unlawful cyber activities.[51] The general purpose of the Convention as laid down in its Preamble is to deter crimes against the confidentiality, integrity and availability of information systems, and the misuse of such systems. The purpose of the Protocol of the Convention on Cybercrime[52] is to supplement the provisions of the Convention on Cybercrime and on the criminalisation of acts of a racist and

---

[50] C. Ernest, Cybercrime: New Threat and Global Response, Expert Group on Cybercrime, Department on New Challenges and Threats, 17–21, January 2011.

[51] C. Wilson, Botnet, Cyber-crime, Cyber Terrorism: Vulnerabilities and Policy Issues for Congress, Congressional Research Service. CRS Report for Congress, 2008, p. 32.

[52] The additional Protocol to the convention on cyber-crime was proposed by some member state. It was the subject of negotiations in 2001 and 2002. Finally, it was adopted by the Council of Europe Committee of Ministers on November 7, 2002. It addressed acts of racist or xenophobic nature committed through computer networks.

xenophobic nature committed through information systems.[53] The Convention has been widely accepted as a landmark, providing for both the substantive and procedural legal frameworks for both domestic and international levels of countermeasures, so as to achieve higher effectiveness in fighting against cyber crimes.[54]

The Convention, which includes 48 articles, starts with the common criminal policy aimed at the protection of society against cyber crime. Criminal acts and the coordination of international law are the primary goals of the Convention. A large part of the Preamble attempts to make human rights a fundamental theme.[55] Article 24 of the Convention on Cybercrime advocates that the parties to the Convention shall cooperate with each other. In addition, it imposes, to the widest extent possible, cooperation regarding criminal offences that are related to computers. Chapter II of the Convention on Cybercrime calls for substantive and procedural measures to be implemented at the national level as well as countering offences against the confidentiality and availability of computer data, such as illegal access, illegal interception, data interference, system interference, misuse of devices respectively in Articles 2, 3, 4, 5, and 6.

In Article 2, the Convention on Cybercrime prohibits unlawful and intentional access, in whole or part, to another computer system. According to the endnote of the Convention, signatory countries are free to define this term in their domestic laws. Similarly, the Convention establishes crimes of illegal interception that must occur 'without right'. However, it protects the internet service provider (ISP) who is entitled to the right to read the communications of its subscribers randomly.[56] However, to remove the ambiguity about who has the right of access to a computer, the drafters have attempted to clarify this point in their explanatory memorandum, citing specific

---

[53] Explanatory Report, Preamble, Convention on Cybercrime, Budapest, 2001.

[54] Convention on Cybercrime, Preamble, (adopted 8 November 2001, entered into force 1 July 2004), paragraph 9.

[55] R. M. F. Baron, 'A critique of the international cybercrime treaty', *Common Law Conspectus* (2002) 10, p. 266.

[56] The Council of Europe Convention on Cybercrime (adopted 8 November 2001, entered into force 1 July 2004).

types of activity that constitute acceptable access and those that are unacceptable.[57]

The most common and prominent method of terrorist attack, which is illegal access, is covered by Article 2 of the Convention on Cybercrime. It covers illegal access when any part of a computer or the whole computer has been accessed without authorised rights. However, this must be committed by "infringing security measures" with the intent of obtaining computer data or with any other dishonest intent. It covers all kinds of technical intrusion and terrorist methods of launching terrorist attacks, especially hacking offences. From a legal point of view, illegal access to computer systems is considered as a "primitive offence" against the integrity and confidentiality of computer data and systems.[58]

According to Articles of the Convention on Cybercrime the misuse of any device used to access or interfere with another's network is forbidden.[59] It states that one cannot produce, sell, or make available a device designed to commit one of the previous five offences that is stated in Chapter 2 of the Convention on Cybercrime. It means a person cannot sell or distribute a device that could "hack" or be used to help another device 'hack' into an unauthorised computer system and consequently, the misuse of devices interpreted as software that have the ability to access the security or hardware of another computer. To prevent related problems, the Convention on Cybercrime drafters have accepted the security analysis. It states that nothing should be interpreted as imposing criminal liability on "authorised testing or protection of a computer system".[60]

The Convention on Cybercrime, in conjunction with other instruments such as the CoE Convention on the Prevention of Terrorism (2005), "can provide sufficient legal basis for cooperation against the

---

[57] The Council of Europe Convention on Cybercrime (adopted 8 November 2001, entered into force 1 July 2004).

[58] P. Brunst, Legal Aspects of Cyber Terrorism, p. 68.

[59] Article 6, The Council of Europe Convention on Cybercrime.

[60] The Council of Europe Convention on Cybercrime (adopted 8 November 2001, entered into force 1 July 2004).

use of the internet for terrorist purposes".[61] Although the Convention on Cybercrime does not define the terms 'cyber crime' and 'cyberterrorism' exactly, Articles 2 through 6 emphasise various forms of criminal activity (illegal access, illegal interception, data interference, system interference, misuse of devices) that are prohibited. Similarly, these activities would fit in with many of the activities that are prohibited in the Convention. During the drafting period of the Convention on Cybercrime, the main concern of the drafters was to confer flexible definitions that have the ability to adapt to new crimes and new technologies in the field of cyber crime. Due to this device, the definitions offer substantive criminal law and procedures applicable to cyberterrorism cases, although some lack of information and procedures still exist within the Convention. In a similar way, the drafters strived to remain within the legal regimes of domestic states. They found it difficult in some areas, because states have different cultures and political views, such as on human rights, data protection, and freedom of speech.[62]

According to Article 22 of the Convention, jurisdiction is granted only on territoriality and nationality theories. Furthermore, the Convention grants authority to the member states to create their own laws and then cooperate with each other in the prosecution of a suspect. The Convention requires that, if an offender is found in a certain state and extradition is required by the injured state, the state that has custody is required to either prosecute the offender or extradite them. Based on territorial jurisdiction a sovereign state has authority to assert jurisdiction over a criminal act that has been committed in its territory. However, according to the ubiquity doctrine, in many cases extraterritorial effect may be applied to the territoriality principle.[63]

---

[61] R. F. Perl, Terrorist Use of the Internet: Threat, Issues, and Options for International Cooperation, Organization for Security and Cooperation in Europe, 2008, p. 5.

[62] S. L. Hopkins, 'Convention on cybercrime: A positive beginning to a long road ahead', *Journal of High Technology Law*, Suffolk University School of Law (2002) 2(1), p. 105.

[63] H. W. Kaspersen, *Discussion paper on cyber-crime and internet jurisdiction*, Council of Europe human rights, 2009, p. 9.

Nevertheless, territorial jurisdiction cannot be implemented against cyberterrorism cases because cyberterrorism operates without respect to any border and ignores the parameters that have been erected by a state law. Furthermore, cyberterrorism occurs in cyber space and its attacks are not only on specific data, but also targeting whole computer systems, infrastructure, and power grids.[64] However, the Convention cannot deter potential cyberterrorism due to the lack of required jurisdiction for cyber terrorist activities. Firstly the territoriality jurisdiction may not be determinable. In other words, it is not clarified where the crime was committed. It is not clear what effect these activities would have on the other states. Secondly, nationality is an elusive issue in today's world.

Therefore, the nationality of most terrorists would not be used in their prosecution because some nations are neither a member of the Convention nor do they care to prosecute or extradite because they support the accuser's actions.[65] The CoE's Convention on Cybercrime was the first step toward an international agreement on cyber activities. Although some of the cyberterrorism offences can be addressed by the Convention on Cybercrime it does not cover and deter cyberterrorism offences specifically. What is more, this Convention has neither been ratified by all EU member states nor has it specifically addressed cyberterrorism.

B. Convention on the Prevention of Terrorism

The CoE has adopted the Convention on the Prevention of Terrorism to increase the effectiveness of existing international texts on the fight against terrorism. CODEXTER drafted the CoE's Convention on the Prevention of Terrorism. The Convention opened for signatories in May 2005 and entered into force in June 2007.[66] The CoE established

---

[64] K. A. Gable, 'Cyber-apocalypse now: Securing the internet against cyber terrorism and using universal jurisdiction as a deterrent', *Vanderbilt Transnational Law* (2010) 43, p. 73.

[65] J. Van de Bograt, International cyber terrorism and jurisdiction, *Florida International University College of Law* (2010), pp. 18–19.

[66] Council of Europe Convention on Prevention of Terrorism (adopted May 2005, entered into force June 2007), Warsaw 16.V. (CECPT).

an intergovernmental CODEXTER matters. It was founded in 2003 to replace the GMT.[67]

The aim of the Convention is to strengthen member states' efforts to prevent terrorism and it sets out two ways to achieve this objective; firstly, it establishes certain acts that constitute criminal offences, namely, public provocation, recruitment, and training. Secondly, it reinforces the cooperation on prevention both internally (national prevention policies), and internationally (modification of existing extradition and mutual assistance arrangements and additional means). What is more, the Convention contains a provision on the protection and compensation of victims of terrorism.

This Convention is a unique one in contrast with other framework decisions of the EU, since it provides the basis of human rights against terrorism.[68] For instance, Article 12 of the Convention requires member parties to adopt and implement anti-criminal measures to be "carried out while respecting human rights obligations, in particular the right to freedom of expression, freedom of association and freedom of religion as set forth in, where applicable to that Party, the Convention for the Protection of Human Rights and Fundamental Freedoms, the International Covenant on Civil and Political Rights, and other obligations under international law".

The laws must be also be "subject to the principle of proportionality, with respect to the legitimate aims pursued and to their necessity in a democratic society, and should exclude any form of arbitrariness or discriminatory or racist treatment". The scope of this Convention is narrower, and it creates new crimes pertaining to terrorism, such as public provocation, recruitment, training for terrorism, and assistance

[67] GMT is a review work of a governmental committee which was drafted following the Convention on the Prevention of Terrorism in 1977 which brought experts together from the 45 Council of Europe member states and observer states. One of the principal functions of the GMT is to make profiles of each member's legislative and institutional counter-terrorism plans. These profiles are short reports on member states' counter-terrorism measures.

[68] K. Nuotio, 'Terrorism as a catalyst for the emergence, harmonisation and reform of criminal law', *Journal of International Criminal Law*, (2006) 45, p. 7.

for victims.[69] It excludes the defence under 'political offences'. According to Article 20, extradition and mutual legal assistance cannot be refused "on the sole ground that it concerns a political offence or an offence connected with a political offence or an offence inspired by political motives".[70]

The CoE's Convention on the Prevention of Terrorism suffers from the lack of general threat provisions with respect to terrorist offences. Therefore it refers itself to the existing treaties such as Convention on Cybercrime.[71] The Convention on Prevention of Terrorism criminalises and prosecutes terrorists, and this can apply to cyberterrorism as well, such as in Article 5 of "public provocation to commit a terrorist offence".[72] What discriminates this treaty from other treaties is it not only bans incitement but also the "public provocation" when the terrorism incident committed causes a danger.[73] Article 5 of the Treaty states:

1. For the purposes of this Convention, 'public provocation to commit a terrorist offence' means the distribution, or otherwise making available, of a message to the public, with the intent to incite the commission of a terrorist offence, where such conduct, whether or not directly advocating terrorist offences, causes a danger that one or more such offences may be committed.
2. Each Party shall adopt such measures as may be necessary to establish public provocation to commit a terrorist offence, as defined in

---

[69] B. David, 'A survey of the effects of counter-terrorism legislation on freedom of the media in Europe', *International Journal of Civil Society Law* (2009) VII(3), p. 38.

[70] Article 20(1), Council of Europe Convention on Prevention of Terrorism (CECPT).

[71] Cyber Terrorism: The Use of the Internet for Terrorist Purposes, p. 75.

[72] M. Gercke *et al.*, 'Terrorist use of the internet and legal response', *Freedom from Fear* (2011). http://www.freedomfromfearmagazine.org/index.php?option=com_content&view=article&id=306 terrorist-use-of-the-Internet-and-legal-response&catid=50:issue-7&Itemid=187 (3 Oct 2001).

[73] B. David, 'A survey of the effects of counter-terrorism legislation on freedom of the media in Europe', p. 38.

paragraph 1, when committed unlawfully and intentionally, as a criminal offence under its domestic law".[74]

Currently, conventional and cyberterrorism offences are not fully covered by any international organisation. The threat of such offences exists, thus the need for a proper response. An additional protocol must also be added to the Convention on the Prevention of Terrorism to cover all illegal terrorist activities, particularly those related to terrorist acts committed in cyber space.

2.3.3 The Group of Eight (G8)

The G8 is an informal forum and it lacks an administrative structure in comparison with other international organisations. The G8 was originally formed in 1975 with six members and it was known as the G6. Canada joined in 1975 and Russia became a formal member in 1998. The leaders have held a summit annually since 1975 to discuss issues of importance, including crime and terrorism, the information highway, economic and political problems.[75] The eight members of the G8 are the US, the UK, France, Germany, Japan, Canada, Italy, and Russia. The ninth member is the CoE but it cannot chair a meeting. The responsibility of hosting rotates throughout the summit cycle at the end of the calendar year.[76] In 1996, the ministers from the G8 summit in France stated:

> States should review their law in order to ensure that abuses of modern technology that are deserving of criminal sanctions are criminalized and the problems with respect to jurisdiction, enforcement powers, investigation, training, crime prevention and international cooperation are adequately addressed. States are urged to negotiate bilateral or multilateral agreements to address the problems of technological crime investigation.

---

[74] Article 5 Council of Europe Convention on Prevention of Terrorism (CECPT).

[75] L. Bantekas *et al.*, *International Criminal Law*, 3rd Edn, Routledge-Cavendish Publisher, United Kingdom, 2007, p. 266.

[76] What is the G8. Available at: http://www.g7.utoronto.ca/ (10 May 2010).

The present study shows the problem of this group in addressing cyber crime and cyberterrorism. This is the reason why they have asked from their member parties to synchronise their legislations. The problem of these organisations is that they do not have harmonised laws, making the prosecution of offenses difficult. Every state willing to deploy their own laws must contribute these laws to other jurisdictions to protect their people.

Consequently, in 1997 at the Denver Summit, the G8 leaders focused on the "investigation, prosecution, and punishment of high-tech criminals, such as those tampering with computer and telecommunication technology, across international borders".[77] It devised 40 recommendations to "increase the efficacy of collective action against transnational organised crime via two goals: strengthening the investigation and prosecution of high-tech crime and more effective regimes for cross-border cooperation in criminal matters".[78]

In 1997, the G08 countries adopted 10 principles to combat computer crime at a meeting in Washington DC. The goal of this meeting was to ensure that no criminal receives "safe haven" anywhere in the world.[79] To combat transnational terrorism and transnational organised crime, an international group of computer crime experts, the G8 Subgroup on High-Tech Crime, was created in 1977, which has promulgated principles and provides best practices regarding the prevention, investigation, and prosecution of computer crimes. The Subgroup also maintains a network of computer crime experts from 50 countries who are available 24 hours a day.[80] These organisations have attempted to strengthen domestic laws to combat terrorists and criminal uses of the internet. Such organisations cooperate with other

---

[77] L. Bantekas *et al.*, *International Criminal Law*, p. 267.

[78] R. Broadhurst, 'Developments in the global law enforcement of Cyber-crime', *An International Journal of Police Strategies and Management* (2006) 29(6), p. 423. Available at: http://dx.doi.org/10.1108/13639510610684674. (14 April 2012).

[79] S. Schjolberg *et al.*, *Harmonizing National Legal Approaches On Cyber-crime, WSIS Thematic Meeting on Cyber Security*, International Telecommunication Union, Geneva, 2005, p. 7.

[80] T. M. Hinnen, 'The cyber front in the war on terrorism: Curbing terrorist use of the internet', *Columbia Science and Technology Law Review* (2008) 5, p. 10.

international partners to secure investigated data and participate in emergency situations when immediate international assistance is required. Participating countries have a dedicated computer crime expert and a means of contact 24 hours a day. The Cybercrime Convention of the CoE requires all parties to establish a 24-hour point of contact for cyber crime cases. International 24-hour response capabilities are thus expected to continue increasing. The G8 and CoE lists will be consolidated.[81]

The G8 Summit has consistently dealt with macroeconomic management, international trade, and relations with developing countries. The concern of this summit is about East–West economic relations whereby energy and terrorism have been of recurrent concern. From this initial foundation, the summit agenda has broadened considerably to include microeconomic issues such as employment and the information highway, transnational issues such as the environment, crime and drugs, and a host of political-security issues ranging from human rights through regional security to arms control.[82] The G8 also held a meeting to test the limits of data protection and privacy laws with a plan to pool terrorism research. This plan was implemented by imposing privacy and security laws in individual countries.[83] The efforts of the G8 in this field have led to countries upgrading their privacy laws to properly detect cyberterrorism cases.

The G8 meetings provide a significant occasion for leaders to discuss major, often complicated, international issues, and to develop personal relations that help them respond in effective and collective fashion to sudden crises or shocks. It gives direction to the international community by setting priorities, defining new issues, and providing guidance to established international organisations.[84]

---

[81] H. M. Jarret *et al.*, *Seizing Computers and Obtaining Electronic Evidence in Criminal Investigations*, 3rd Edn, Office of Legal Education Executive Office for United States Attorneys Publication, p. 58.

[82] What is the G8. Available at: http://www.g7.utoronto.ca/ (10 May 2010).

[83] G8 pool data on terrorism. Available at: http://www.guardian.co.uk/uk/2005/jun/18/g8.usa (10 May 2010).

[84] What is the G8. Available at: http://www.g7.utoronto.ca/what_is_g8.html (10 May 2010).

The Financial Action Task Force (FATF) was established at the 1989 G7 Summit in Paris in response to mounting concern over international money laundering. Following the 11 September 2001 terrorist attacks, a meeting on the financing of terrorism was held in Washington DC and it was decided to expand the FATF mission to include combating terrorist financing.

In 2007, the G8 specifically expressed its support for the FATF, promising in the Heiligendamm Statement on Counter-Terrorism to implement and promote all 40 of the FATF's Recommendations on Money Laundering and the Nine Special Recommendations on Terror Financing.[85] FATF recommendation grouped into three categories (criminal law, banking law, and international cooperation) to cover all offensive activities. The main problem of the FATF in combating cyberterrorism and money laundering is the gap that exists among its members. FATF established a peer review process that including two parts to assess has implementation of 40 recommendations. First, each FATF member annually engages in self-assessment through the use of a standard questionnaire. Second, members are requires to undergo a periodic mutual evaluation process that involves a site visit by three or four experts from other member governments.[86]

The leaders of the G8, in a meeting on 16 July 2006 in St. Petersburg, condemned terrorist attacks and expressed their deepest sympathy and denounce the G8 summit on counter-terrorism. They decided to enhance cooperation among themselves. They also declared their priorities and commitment to combat terrorism. Significant efforts included:

(1) Implementing and improving the international legal framework on counter-terrorism;
(2) Ensuring national legislation is adapted, as appropriate, to address new terrorist challenges;
(3) Suppressing attempts by terrorists to gain access to weapons and other means of mass destruction;

---

[85] Compliance report of group G8 2008. Available at: http://www.g8.utoronto.ca/evaluations/2008compliance-final/18-08-terrorism.pdf (11 May 2010).

[86] R. Braodhurst, 'Development in the global law enforcement of cyber-crime', p. 453.

(4) Engaging in active dialogue with civil society to help prevent terrorism;
(5) Enhancing efforts to counter the financing of terrorism based on agreed standards;
(6) Developing and implementing an effective strategy to counter terrorist propaganda and recruitment, including with regard to the use of suicide bombers;
(7) Effectively countering attempts to misuse cyber space for terrorist purposes, including incitement to commit terrorist acts, to communicate and plan terrorist acts, as well as recruitment and training of terrorists;
(8) Preventing any abuse of the migration regime for terrorist purposes while at the same time facilitating legitimate travel;
(9) Bringing to justice, in accordance with obligations under international law, those guilty of terrorist acts, as well as their sponsors, supporters, those who plan such acts and those who incite terrorist acts;
(10) Ensuring and promoting respect for international law, including international human rights law, refugee law and humanitarian law in all our counter-terrorism efforts;
(11) Promoting supply chain security, based on existing international standards and best practices; and,
(12) Promoting international cooperation in subway, rail and road security and in raising standards in aviation, and maritime security.[87]

2.3.4 Asia-Pacific Economic Cooperation (APEC)

The APEC forum is a regional forum which was established in 1989 for facilitating economic growth and its goal is to strengthen the Asia-Pacific community. APEC has 21 members and decisions made within APEC are reached by consensus. After the September 2001 terrorist attacks on the US, APEC issued a statement on counter-terrorism and

[87] G8 Summit Declaration on Counter Terrorism, G8 Information Centre. Available at: http://www.g8.utoronto.ca /summit/2006stpetersburg/counterterrorism.html (10 May 2010).

condemned these attacks and increased its efforts to collaborate to fight against terrorism.[88]

The Telecommunications and Information Ministers of the APEC economies issued the Statement on the Security of Information and Communications Infrastructures and a Program of Action in 2002, supporting measures taken by members to fight against misuse of information. They designed six recommendations as the basis for APEC's fight against cyber crime comprising legal development, information sharing and cooperation, security, and technical guidelines, public awareness, training and education, and wireless security.

The main objective of the APEC is to enhance cooperation and harmonisation of high level standards. APEC also attempts to expand cooperation among its member states in combating cyberterrorism. This organisation has started engaging in serious and remarkably rapid institutional changes in response to terrorism activities after the September 11 attacks. APEC began to implement its action plans on security matters and counter-terrorism in November 2001. The hurdles that APEC is confronting in combating cyberterrorism offenses are cooperative preventive measures, cross-border sharing of information, and recovery management.

In 2003, a survey of laws was carried out by the E-Security Task Group. Following a meeting that was held in Bangkok in 2003, the objectives of the meeting were stated to be to assist economies to develop the necessary legal frameworks to promote the development of law enforcement capacity and to strengthen cooperation between private and public sectors in addressing the threat of cyber crime. At the conference, the experts present agreed that every economy needed a legal framework, including one for substantive and procedural law and one for the law and policies of inter-economies cooperation. They established the role of international instruments, particularly the Convention on Cybercrime. They also emphasised jurisdictional cooperation, law enforcement construction, and the capacity building of the investigators.

---

[88] R. Feinberg, 'Voluntary multilateralism and institutional modification: The first two decades of Asia pacific economic cooperation (APEC)', *The Review of International Organizations* (2008) 3, pp. 239–258.

The sixth APEC Ministerial Meeting on the Telecommunications and Information Industry made a declaration encouraging all economies to enact a comprehensive set of laws relating to cyber security and cyber crime that are consistent with international legal instruments, including UN General Assembly Resolution 55/63(2000) and the Convention on Cybercrime (2001). However, economies cannot enact a unified legal instrument, because there are many differences between the economies of APEC members.[89]

2.3.5 North Atlantic Treaty Organization (NATO)

The NATO was founded in 1949 on the basis of the principle of collective defence. The parties followed the basic principle of the Charter of the UN. This cooperation and commitment to security continues among the member nations today. The North Atlantic Treaty has established NATO as a collective security and defence unit pertaining to cooperation in terms of security and defence policies as well as military operations.[90] Nevertheless, the internet, cyber space and cyber crime did not exist at the time that NATO was established.

However, NATO has adapted to change its political and technical requirements and advance its capabilities in the area of cyber defence. It addresses new challenges that are posed by terrorists and the threats to computer information systems (CIS).[91] It has recently taken steps toward a common policy on cyber defence in 2008 to protect critical information systems and strengthen international and national cooperation.

Despite its many obligations, NATO does not determine specific military action on cyber attacks. NATO does not extend its collective

---

[89] L. Xingan, 'International actions against cybercrime: Networking legal systems in the networked crime scene'.

[90] U. Haubler, Cyber Security and Defence from the Perspective of Articles 4 and 5 of the NATO Treaty. In: E.Tikk and A.-M. Taliharm (Eds.), *International Cyber Security Legal Policy Proceeding, CCD Coe*, 2010, p. 102.

[91] P. Everard, Nato and cyber terrorism, in Centre of Excellence Defense against Terrorism, *Responses to Cyber Terrorism*, Ankara, Turkey, 2008, p. 125.

self-defence policy to the target country of cyber attacks.[92] Cyber defence falls under NATO's jurisdiction when all members agree that a certain issue should. For the time being, NATO action is limited to active defence and does not engage in active cyber offense. NATO's dependence on IT brings a high level of risk. This organisation faces the same global threats as other institutions. Although protecting its members from cyberterrorism is a challenge for NATO, it has established bodies to protect critical infrastructure with initial operating capability. Upgrading their systems to detect and respond to cyber attacks and train experts in computer and network security were that is difficult for NATO. NATO must provide anti-terrorism malware support, comprehensive vulnerability notification, and utilise mail content checking, and monitor intrusion detection system.[93]

NATO is the only military organisation that has the capability to deter and defend against cyberterrorism. The reality of cyber attacks is that they are becoming more disastrous, frequent and costly, and threaten nations' critical infrastructures. They have the potential to threaten national security and stability. Therefore, NATO proposes a strategy to enhance national defence capabilities by combining cyber deterrence with a central defence system and bringing all its members under central cyber protection. This strategy marked the first time in an international organisation that members were required to cooperate in their national cyber defence capabilities.[94] The possibility of a large-scale attack that includes military force is more than a mere missile attack. Attackers also need adequate time and proper strategies to react to their target's response to the initial missile attack. Responding to cyber attacks that cause extensive destruction may require a shorter reaction time.

NATO's emphasis on responding to cyber attacks is the protection, prevention, and recovery of its command and control systems or

---

[92] K. A. Gable, 'Cyber apocalypse now: Securing the internet against cyber terrorism and using universal jurisdiction as a deterrent', p. 34

[93] P. Everard, Nato and cyber terrorism, p. 125.

[94] M. Dugrol, 'Developing an international cooperation on cyber Defense and deterrence against cyber terrorism', p. 12.

energy grids. According to several experts, NATO is expecting attacks from cyber space after receiving unconventional attacks in recent years. According to a report from experts, due to the fact that attacks on NATO in recent years were unconventional, the next expected attack on NATO might be expected from cyber space.

Cyber attacks have become more organised, more frequent, and cyber terrorists are becoming motivated to carry out their attacks via cyber space, NATO aims to combine a cyber deterrent strategy with a strong defence system. Since it ascertained that it could be a target, NATO has asked its members to coordinate and cooperate to enhance national cyber defence and to bring bodies under the same, centralised cyber protection process.[95] Going by unprecedented threats targeted at NATO in recent years, which were unconventional, it is anticipated that a future attack might be received from cyber space. Therefore, NATO has stepped up efforts toward establishing assisting bodies such as the Cyber Defence Management Authority (CDMA), the Cooperative Cyber Defence Centre of Excellence (CCD CoE), and the Computer Incident Response Capability under NATO C3.[96]

NATO has established the NATO Computer Incident Response Capability (NCIRC) — created following the events of 11 September 2001 — providing intrusion detection and response capabilities, malicious code prevention, incident handling, computer forensics, vulnerability assessment, and security configuration support to NATO under NATO C3.[97] The complexity of cyber defence requires a wide range of policies to prepare for, prevent, detect, respond to, recover, and learn lessons from cyber attacks. Although implementing defence mechanisms is not easy, action must be swift and immediate because cyber attacks unfold in nanoseconds. NATO C3 established and oversees a CDMA to coordinate immediate and effective cyber defence action. The agency can organise and dispatch rapid reaction teams to

[95] M. Dugrol *et al.*, 'Developing an International Cooperation on Cyber Defense and Deterrence Against Cyber Terrorism', p. 12.

[96] M. Dugrol *et al.*, Developing an international cooperation on cyber defense and deterrence against cyber terrorism, p. 12.

[97] P. Everard, Nato and cyber terrorism, p. 125.

support its allies in defending against cyber attacks.[98] The NATO C3 agency is responsible for executing NATO's political and military mission. This agency also has a strong role in cyber defence and in the initial selection and assessment of security tools.

In 2008, following the approval of the NATO Policy on Cyber Defence, NATO established a NATO CDMA. The NATO CDMA initiates and coordinates immediate and effective cyber defence actions. The Policy on Cyber Defence establishes the basic principles and provides direction for NATO civil and military bodies and recommendations to NATO nations in order to ensure a common and coordinated approach to cyber defence and response to cyber attacks.

To facilitate cooperation and assistance between the NATO CDMA and cyber defence organisations in NATO member nations, a Memorandum of Understanding/Agreement (MOU/MOA) needed to be established. The MOU/MOA sets out the rules and procedures for the exchange of information and services between both parties, including pre-arrangements for assistance during a cyber attack.[99] Then NATO established a CCD CoE, which is hosted by Estonia, on 28 October 2008.[100] The mission of the CCD CoE is to enhance the cooperative cyber defence capability of NATO in order to support NATO's military transformation and to sponsor nations and other customers, thus improving NATO's interoperability in the field of cooperative cyber defence. It is an international military organisation and it is not part of the NATO command structure. Participation in this Centre of Excellence is open to all NATO members.

However, the Centre of Excellence encountered challenges in cooperative defence against cyber attacks. This body must first support the infrastructure and administrative bodies of each nation. In fact, allocating the budget among countries in a fair manner is difficult. The annual budget is approved according to the shared cost agreed in

---

[98] NATO Getting Serious about Cyber Security, 17 November 2011. Available at: http://www.defencemanagement.com/feature_story.asp?id=18166 (12 July 2013).
[99] B. Bottesini, International Cooperation in the Fight against Cyber Terrorism. Available at: http://www.ata-sac.org/index.php?lang=en (23 May 2010).
[100] P. Everard, Nato and cyber terrorism, p. 125.

the MOU by the committee. The committee must find new sponsoring nations that will contribute participants and manage the necessary budget to enable the nations to defend against cyber attacks. The exchange of information among countries is also another difficulty that may confront countries because of confidentiality issues.

The CCD CoE's mission is to enhance the capability, cooperation, and information sharing among NATO members. This organisation has attempted establish this kind of cooperation through consultation, education, research, and development. CCD CoE must study all of its members' regulations to implement legal policies that need large amounts of money and expertise. It must find and understand technical solutions to secure deployed systems. Detecting technical attacks and providing offensive, defensive, and exploitative strategies is also difficult.[101]

The collective security and defence mechanisms provide a lawful and legitimate manner in tackling the threats as stated in Articles 4 and 5 of the North Atlantic Treaty. Article 4 provides that: "the Parties will consult together whenever, in the opinion of any of them, the territorial integrity, political independence, or security of any of the Parties is threatened".

Article 5 of the Treaty indicates that each party is its own judge and has the prerogative to determine if armed attack is a sufficient response, or on any other action. NATO's response to the September 11 attacks on the US is an instance of the application of Article 5 of the North Atlantic Treaty. Article 51 of the UN Charter[102] and Article 5 of the North Atlantic Treaty have significant terminological over-

[101] Cooperative cyber defence centre of excellence (CCD CoE), Talline, Estonia. Available at: https://www.ccdcoe.org/ (26 June 2013).

[102] Article 51 of the United Nations Charter states: "Nothing in the present Charter shall impair the inherent right of individual or collective self-defence if an armed attack occurs against a Member of the United Nations, until the Security Council has taken measures necessary to maintain international peace and security. Measures taken by Members in the exercise of this right of self-defence shall be immediately reported to the Security Council and shall not in any way affect the authority and responsibility of the Security Council under the present Charter to take at any time such action as it deems necessary in order to maintain or restore international peace and security".

laps. Both the UN and NATO shared the same views on the September 11 attacks, that these attacks were included in Article 5 of the North Atlantic Treaty and Article 51 of UN Charter. Nevertheless, none of them explains and determines why the armed attack is applicable to such an attack. (The provisions do not explain if armed attack is a suitable response to such an attack that threatens the security of a nation.)

NATO can employ security measures, when a certain amount of damage is caused to the cyber infrastructure of one of its member states. The attacks were so grave and widespread that it was considered whether Article 5 of the NATO should be invoked. As it is mentioned above, Article 5 states that an assault on one allied country obligates the alliance to attack the aggressor.[103] The case of cyber attack upon Estonia exposed several realities. Firstly, they revealed that cyber attacks were extremely difficult to trace, thus making them attractive to terrorist groups or a government which desires to remain anonymous. Secondly, the attacks were relatively easy to carry out (in fact, many users were not even aware that their computers were used in the attack). Thirdly, the Estonian incident showed that cyber attacks could cause real world harm and result in significant costs and chaos.[104]

In the above attack, the terrorists' followers started the attack by themselves. Then the hackers and "script kiddies" were incited to launch DoS attacks against Estonia in chat rooms.[105] Therefore, Estonia requested emergency assistance to defend its digital infrastructure against the cyber attacks it received, as a member of NATO. The Estonian Minister of Defence, Jaak Aaviksoo, stated that the "the attacks were aimed at the essential electronic infrastructure of Estonia

---

[103] Joshua Davis, 'Hackers take down the most wired country in Europe', *Wired Magazine*, Aug 21, 2007, Available at: http://www.wired.com/politics/security/magazine/15-09/ff (13 July 2012).

[104] W. McGavran, 'Intended consequences: Regulating Cyber-attacks', *Tulane Journal of Technology and Intellectual Property* (2009) 12, p. 270.

[105] P. Brunst, Legal Aspects of Cyber Terrorism, p. 54.

and this was the first time that a botnet threatened the security of an entire nation".[106]

Before the Estonian incident, NATO did not view cyber attacks as military actions. It is clear that NATO lacked "both a coherent cyber doctrine and a comprehensive cyber strategy". NATO's Cyber Defence Program lacked the actual need to counter cyberterrorism. Therefore, NATO did not include these attacks under the fifth Article of the NATO agreement on military protection. The applicability of Articles 4 and 5 of the North Atlantic Treaty and the response threshold for a cyber attack must be assessed on a case-by-case basis. These Articles "indicate that it is the nation's prerogative to determine whether they consider themselves exposed to a threat or under an armed attack. However, they do not create any automatism whatsoever concerning the response in such case".

NATO lacked the necessary capabilities against cyber attacks, which had gradually became a tangible security threat to the communication systems of NATO. The cyber attack against Estonia revealed NATO's cyber defence inadequacies, forcing the alliance to reconsider its defence strategy. NATO was alerted to the Estonia attack in 2007 when the country's computer networks were paralysed by a cyber attack blamed on Russia.[107] Cyber attacks rarely have a direct military dimension. Thus, the alliance may actually have a limited role in dealing with such attacks. NATO members have recently agreed to strengthen its cyber defence mechanism. NATO Secretary General Anders Fogh Rasmussen emphasised that the cyber defence of NATO must be reviewed because they have received about 2,500 significant cyber attack cases in their system. Rasmussen, in a news conference, said that, "We are all closely connected, so an attack on one ally, if it is not dealt with quickly and effectively, can affect us all. Cyber

---

[106] V. Joubert, Five Years after Estonia's Cyber-attacks: Lessons Learned for NATO, Research Division — NATO Defence College, May 2012, p. 2.

[107] Adrian Croft, 'NATO boosts cyber defenses but members differ on its role', yahoo news, 4 June 2013. Available at: http://news.yahoo.com/nato-boosts-cyber-defenses-members-differ-role-205618606.html (7 June 2013).

defence is only as effective as the weakest link in the chain. By working together we strengthen the chain".

However, larger countries such as the US, UK, France, and Germany disagree because they are already spending large sums on cyber defence within their own territories. These countries are reluctant to divert their own money to NATO activities that are expected to primarily benefit others. The alternative offered by Rasmussen is to use the rapid reaction teams of NATO to aid countries that become targets of cyber attacks.[108]

A decision on whether to utilise Articles 4 or 5 of the Treaty will thus "depend on political policy perceptions, and the different roles played by the government agencies involved on the examination and assessment of cyber threats and incidents, and competent to adopt or contribute to actual response".[109] Another legal issue that arises for NATO cyber defence is the applicability of existing humanitarian law to cyber attacks. There are two approaches among scholars. Some of them believe in the "consequence-based approach" which interprets the notions of "armed conflict" and "attack" in the cyber domain, while others believe in the "actor-based approach". The adoption of the "consequence-based approach" would make *jus in bello* applicable to cyber attacks and reflect the reality of modern warfare.

This is because the potential of cyber attacks for causing damage, death, injury or destruction can be similar to physical attacks or incidences of armed conflict from a humanitarian perspective and thereby outlawed.[110] "Estonia felt so helpless in the face of DDoS attacks on its networks it asked NATO for military assistance. Even the mighty Google had to concede a need for help with Operation Aurora,

---

[108] Adrian Croft, 'NATO boosts cyber defenses but members differ on its role'.

[109] U. Haubler, Cyber Security and Defence from the Perspective of Articles 4 and 5 of the NATO Treaty, *The 2010 Conference on Cyber Conflict Law and Policy Track, International Cyber Security Legal and Policy Proceedings, CCD CoE*, Tallin, Estonia, 2010, p. 276.

[110] V. Joubert, Five Years after Estonia's Cyber- attacks: Lessons Learned for NATO, p. 3.

contracting with the NSA for assistance".[111] The Estonian Justice Minister asked the EU to include this act as an act of cyberterrorism.[112]

By analysing Article 5 of the North Atlantic Treaty, which is the collective self-defence obligation in case of armed attack, it is concluded that it would be fully compatible with Article 51 of the UN Charter, since Article 51 of the UN Charter reveals similarly expressed intentions about "armed attack". According to the analysis of scholars such as Beckett and Kaplan, when Article 5 expresses that an armed attack against one or more of the Parties shall be considered to be an attack against them all, it fully expresses what the inherent right of collective self-defence means.[113] NATO stepped forward to maintain the international peace and security within the framework of the UN Charter based on its Article 51. However, Article 7 of the North Atlantic Treaty recognises "the primary responsibility of the Security Council for the maintenance of international peace and security". In addition, in Article 5, it is stated that "any measures taken by the NATO countries as the result of an armed attack upon them shall be terminated when the Security Council has taken the measures necessary to restore and maintain international peace and security".[114]

An overlap exists in terms of looming security threats between Article 35 of the UN Charter and Article 4 of the North Atlantic Treaty. Thus, if "the consultation leads to the conclusion that the threat is sufficiently serious, one or other or all of the parties will exercise the right which they have under the Charter to bring the matter before the Security Council".[115] The regulations on armed attacks can apply to cyberterrorism as well because cyberterrorism can be considered a kind

---

[111] D. B. Hollis, 'An E-sos for cyberspace', *Harvard International Law Journal* (2011) 52, p. 389.

[112] S. Myrli, Nato and cyber defence, 173 DSCFC 09 E bis, Nato Parliamentary Assembly, Annual Committee Report 2009. Available at: http://www.nato-pa.int/default.Asp?SHORTCUT=1782 (23 Mar 2013).

[113] L. K. Kaplan, *Nato 1948, The Birth of Transatlantic Alliance*, Rowman & Littlefield Publishe, United States of America, 2007, pp. 230–235.

[114] The North Atlantic Treaty (adopted in 1949, came into force in), (NATO) Art 5.

[115] U. Haubler, Cyber Security and Defence from the Perspective of Articles 4 and 5 of the NATO Treaty, p. 276.

of armed attack. Armed attack regulation in international law has evolved to make states responsible for providing support against persons involved in terrorist activities. As previously mentioned, one option for responding to cyberterrorism offenses is using current regulation on terrorism activities. Armed attack regulation is also applicable to cyberterrorism regarding attribution and self-defence.

2.3.6 International Multilateral Partnership against Cyber Terrorism (IMPACT)

This is the international mechanism in dealing with cyberterrorism. During the ITU Telecom World 2009 Fair in Geneva, the chief of Union's telecommunication agency warned that the next world war could happen in cyber space. In addition his secretary added that "loss of vital networks would quickly cripple any nation, and none is immune to cyber-attack". In recent years, by linking the internet with daily lives, cyber attacks and crimes have increased. Therefore individual countries have started to respond by bolstering their defences. Countries such as the US, South Korea, and the Baltic state of Estonia have attempted to confront cyber attacks as they are the most common victims of cyber attacks.

The IMPACT, backed by the UN ITU and the International Criminal Police Organization (Interpol), is known as the world's first global public–private partnership against cyber threats and launched its global headquarters in Cyberjaya, Malaysia on 20 March 2009.[116]

On 17 May 2007, the ITU launched the Global Cyber-Security Agenda to provide a framework within its international response to the growing challenges to cyber security that could be coordinated and addressed. The ITU and IMPACT formally gathered in IMPACT's new state-of-the-art global headquarters in Cyberjaya. Established in September 2008, IMPACT is the first comprehensive global partnership between governments, industry leaders, and cyber security experts to enhance the global community's capacity to prevent, defend, and respond to cyber threats.

---

[116] Available at: http://www.impact-alliance.org/ (23 May 2010).

IMPACT seeks to bridge the gap that exists between domestic and international spheres in countering cyber threats. It promotes greater cooperation in combating cyber threats. IMPACT is backed by and collaborates with the ITU and it is addressed as the operational home for the ITU. The ITU is built upon five strategic pillars and made up of seven main strategic goals. It seeks to undertake legal, technical and procedural measures, and organisational structures at the national and regional levels with regards to harmonisation at the international level. This cooperation is intended to develop a framework for international cooperation, dialogue, and coordination in dealing with cyber threats. This information society is borderless. So the response mechanism related to cyber criminal activities is not adequate.[117]

The IMPACT nerve-centre in Cyberjaya has a built-up area of 58,000 square feet on a 7.8-acre piece of land. The Malaysian government has provided a grant of US$13 million (about RM43 million) as initial funding to build the facilities. This IMPACT HQ will host the ITU's Global Cyber-Security Agenda, which promotes international cooperation to make cyber space more secure in an increasingly networked information society. IMPACT has four key divisions:

(i) Global response centre;
(ii) Training and skills development centre;
(iii) Centre for security assurance and research; and,
(iv) Centre for policy and international cooperation.

It will act as a centralised anti-cyberterrorism intelligence centre which allows its 191 member countries to be alerted to cyberterrorism threats such as attacks against the global financial system, power grids, nuclear plants, and air traffic control systems.[118] Former Prime Minister of Malaysia Datuk Seri Abdullah Ahmad Badawi said IMPACT was the direct result of the need for a diverse range of stakeholders to come together to handle the challenges of cyber disruption.

---

[117] Wikipedia Encyclopedia Available at: http://en.Wikipedia.org/wiki. (20 Aug 2009).
[118] Available at: http://www.impact-alliance.org/ (23 May 2010).

IMPACT is a centre that cooperates and collaborates with stakeholders encompassing government and private institutions to provide sufficient security for combating cyberterrorism. The main aim of IMPACT is to secure cyber space. As a rapidly emerging threat for governments and businesses, cyber security has enjoyed rising stature and attention, but the classification of threats has curtailed the possibility for cooperation among nations and between business and government. Cyber security is challenge beset with complexity because it transcends and cover numerous aspects. The connected nature of the internet has made cyber security a unique international challenge. Given that cyber infrastructure is heavily concentrated within the private sector, cyber security threats present challenges to both the public and the private sectors. The increasing dependence of individuals on the internet also indicates that cyber security should be integrated into the balance between national security and individual freedom. IMPACT collaborates with the public and private sectors with specialised tools and systems such as the Electronically Secure Collaborative Application Platform for Experts[119] to enable partner countries and the global community to respond immediately to cyber threats, especially during crises.[120]

2.3.7 The Organisation for Economic Cooperation and Development (OECD)

The OECD is a unique forum where the governments of 30 democracies work together to address the economic, social and environmental challenges of globalisation. The OECD has been working for many

---

[119] ESCAPE is a unique electronic tool that enables authorized cyber experts across different countries and verticals to pool resources together to collaborate with each other remotely, within a secured environment. These individuals may include IT experts, regulators, Computer Incident Response Teams (CIRTs), and white hats. By converging resources and expertise from many different countries at short notice, ESCAPE enables partner countries and the global community to respond immediately to cyber threats, especially during crises.

[120] Available at: http://www.impact-alliance.org/home/index.html (20 May 2012).

years on a range of policy issues associated with the information society. These include infrastructure and services, consumer protection, privacy and security, through to broader issues surrounding ICT and economic growth.[121]

In 1983, the OECD established an expert committee to discuss the computer crime phenomenon and criminal law reform. Some of the offences listed in the 1985 OECD document are unauthorised access, damage to computer data or computer programmes, computer sabotage, unauthorised interception, and computer espionage. It defines computer crime and computer-related crime as "any illegal, unethical, or unauthorised behavior involving automatic data processing and/or transmission of data". This is "including unethical behavior within the criminal definition without more amplification would likely be struck down as unconstitutionally vague".[122]

The OECD and its committee is one of the leading organisations and international efforts dealing with the criminal law issues relate to cyber crime. The OECD's committee presented its recommendation in 1985, that due to the nature of cyber crime, international cooperation was needed to reduce and control such activities and recommended to member countries to change their penal legislation in order to include cyber crime in their legislation.[123] The committee of experts proposed a guideline for national legislators as well as a list of offences to enable uniform criminal policies on legislation which reflected the general consensus on computer-related crimes as pertaining to criminal law from 1986 to 1989. The enumerated computer crimes which have been achieved include "computer fraud, computer forgery, damage to computer data or computer programs, computer sabotage, unauthorized access, unauthorized interception,

[121] Available at: http://www.intgovforum.org/brief.htm (2 Jun 2010).

[122] R. W. Aldrich, 'Cyber terrorism and computer crime issues surrounding the establishment of an international legal regime', *USAF Institute for National Security Studies* (2000), p. 22.

[123] Z. K. Shalhoub, *Cyber Law and Cyber Security in Developing and Emerging Economies*, Edward Elgar Publishing, United Kingdom, 2010, p. 2.

unauthorized reproduction of a protected computer program and unauthorized reproduction of a topography".[124]

The OECD developed the Guidelines for the Security of Information Systems and Networks in July 2002, calling on member governments to "establish a heightened priority for security planning and management", and to "promote a culture of security among all participants as a means of protecting information systems and networks".[125] The aim of these guidelines is to develop a "global culture of security" through advice on policies and measures that address internal and external threats such as cyberterrorism, computer viruses, and hacking in a globally interconnected society, while preserving important societal values such as privacy and individual freedom. Despite being non-binding, the recommendations of these guidelines reflect a consensus among prominent jurisdictions that the online environment affects security.[126]

2.3.8 The Association of Southeast Asian Nations (ASEAN)

The ASEAN operates on a Pan-Asian approach.[127] It includes ten nations: Brunei Darussalam, Cambodia, Indonesia, Laos, Malaysia, the Philippines, Singapore, Thailand, Myanmar, and Vietnam. It forms a regional forum for matters of Mutual Legal Assistance (MLA). The methodology of ASEAN mirrors the approach of the EU. ASEAN conducted four ministerial meetings on transnational crime in Manila 1997, Yangon 1999, Singapore 2001, and Bangkok 2003.

The focus of these meetings was on transnational crime and cooperative efforts in combating such crimes. The first meeting held in Manila discussed transnational crime and issued a declaration aimed at regional cooperation on criminal matters. The declaration proposed

---

[124] Global Cyber Law Database, International Development of Cyber Law, Available at: http://www.cyberlawdb.com/main/international-development (12 Nov 2011).

[125] Available at: http://www.intgovforum.org/brief.htm (2 Jun 2010).

[126] Global Cyber Law Database, International Development of Cyber Law, Available at: http://www.cyberlawdb.com/main/international-development (12 Nov 2011).

[127] Pan-Asian approach is an ideology that promotes the unity of Asian peoples.

substantial enhancements in regional law enforcement cooperation. ASEAN ministers agreed to have biannual meetings for the following agendas: coordination of the relevant bodies' activities, signing of MLA treaties and other treaties among ASEAN member countries, establishment of an ASEAN Center for Transnational Crime to coordinate regional efforts against international crime such as cyberterrorism, creation of an ad hoc expert group to develop action plan for tackling transnational crime for regional cooperation, and facilitation of cooperation among law enforcement by encouraging its members.[128] The next meeting in Yangon also dealt with issues on collective efforts against organised crime. In the third and fourth meetings, the ministers committed to collaborating further to combat computer-related crime and called for a partnership between ASEAN and other agencies such as Interpol and the UN.[129] The capacity of this MLA to effectively combat transnational crime such as cyber crime and cyberterrorism is illustrated by the action plan of the ASEAN and the China Cooperative Operations in Response to Dangerous Drugs, in partnership with the UN Drug Control Program.

2.4 BILATERAL LEVEL OF EFFORT

There is a necessity for a multilateral effort to combat cyberterrorism offences, due to different countries' rules for extradition and legal assistance as well as various laws governing computer crime. Bilateral efforts to prevent and respond to cyber attacks have advantages and disadvantages. The first obstacle is the scope of mutual legal assistance treaties (MLATs) which are narrow in terms of the number of countries each can involve. Another problem is that, most of these treaties do not cover cyberterrorism in specific manner. The final problem is due to the significance of time which plays a crucial role in cyberterrorism offences and the application of MLATs is time-consuming because it involves more paper work and bureaucratic procedures. MLATs may advance the security and safety goals of countries. They allow

---

[128] ASEAN Declaration on Transnational Crime, signed on 20 December 1997 in Manila. Available at: http://www.aseansec.org/politics/adtc97.htm (31 Feb 2013).
[129] R. Broadhurst, 'Developments in the global law enforcement of cyber-crime', p. 424.

extradition and prosecution which will significantly enhance deterrence of cyberterrorism, because with such treaties the difficulties in jurisdiction investigation will be removed. They will encourage cooperation among signatory countries. Furthermore, they can enhance technical cooperation and go beyond the limitations and confines of the treaty.

A good example of such treaty is an agreement that was signed between Estonia and Russia in 1993 to render each other legal assistance and this provided a legal procedural act which the parties implemented, called the Agreement on Mutual Assistance between Estonia and Georgia. In the articles of this agreement some procedural acts are enumerated but they were not exclusive. Following the incidents on 10 May 2007 in Estonia, a letter was submitted by the Estonian public prosecutor's office to the Russian Federation requesting assistance for preliminary investigations. They asked for a procedural activity defined as identification of persons, based on Articles 206, 207, and 208 of the Penal Code, regarding the crimes of computer sabotage, damaging of connections to computer networks, and the spread of computer viruses.

The Russian Federation replied that according to the agreement, such legal assistance was rendered based on the legal acts of the contracting party who had received a request and in this case there was no need for cooperation in the field of operative prosecution measures to find out the location of a person. (Russia has not shown any consistency in interpreting this mutual assistance agreement. They move ahead according to the interpretation that best suits their interests.)[130]

Estonia signed and then ratified the European Convention on Mutual Assistance in Criminal Matters and its protocol on 19th February 1997 and it entered into force on 27th July 1997; also, the second additional protocol was ratified on 9th June 2004 and entered into force on 23rd July 2004. Russia ratified it on 10th December 1999 and it entered into the force in 2000. However, it has not ratified the second additional protocol.

---

[130] T. Eneken *et al.*, Legal Cooperation to Investigate Cyber Incidents: Estonian Case Study and Lessons, *Proceedings of the European Conference on Information Warfare*, Estonia, 2010, p. 288.

According to the Article 2 of the Convention, any refusal of mutual assistance must be based on two criteria: first, if the request concerns an offence which the requested party considers a political offence, an offence connected with a political offence, or a fiscal offence (Russia has identified the criteria to determine the possible crimes that correspond to those characteristics in a declaration made to the convention), or second, if the requested party considers that execution of the request is likely to prejudice the sovereignty, security, public order, or other essential interests of herself as a receiving country.[131]

According to Article 1 of the Convention, the contracting parties undertake to afford each other the widest measure of mutual assistance in proceedings in respect of offences the punishment of which, at the time of the request for assistance, falls within the jurisdiction of the judicial authorities of the requesting party. According to Article 3.1, the requested party shall execute in the manner provided for by its law any letters rogatory relating to a criminal matter and addressed to it by the judicial authorities of the requesting party for the purpose of procuring evidence, among other things.[132]

However, an assessment conducted by the US government and the private sector determined that the cyber attacks were most likely carried out by politically motivated hackers and not by Russians. The Deputy Director of the US Cyber Emergency Response Team (USCERT) surmised that the "zombies" utilised as slave computers by botnets originated in the US and such attacks lacked the hallmarks of the major powers. However, some Russian hackers, such as SpORaw, believed that the attacks on Estonia could not have been implemented without the consent of the Russian authorities. Based on chat room postings and by the fact that on some Estonian sites attackers replaced the website homepage with the phrase "Hacked by Russian hackers", they contended that the hackers were acting under "recommendations" from Russian authorities.[133]

---

[131] Article 2 Convention on Mutual Assistance in Criminal Matters.

[132] Articles 1–3 Convention on Mutual Assistance in Criminal Matters.

[133] S. C. Shackelford, 'From nuclear war to net war: Analogizing cyber-attacks in international law', *Berkley Journal of International Law* (2009) 27, pp. 198–199.

In its answer to the European Commission's inquiry on the subject, the Estonian Ministry of Justice pointed out the following issues with Russia regarding cooperation in criminal matters:

(1) Revision of a letter rogatory generally takes much time and reminders are ignored;
(2) Assistance is refused for procedural activities regarding suspects; this is justified by referring to the fact that the notion of "suspect" does not exist in Russian legislation; also, Russia will not interrogate a person of Russian citizenship;
(3) A prior court ruling is required as a precondition for transferring of documents;
(4) Covert investigation is refused without a court order (in Estonia, the relevant authorisation is issued by the Public Prosecutor's Office);
(5) On occasions, Russia has insisted that a particular request be submitted through Interpol — this was also the case in relation to the letter rogatory concerning the April/May 2007 cyber attack.[134]

2.5 HARMONISATION AND COOPERATION OF INTERNATIONAL ORGANISATIONS

As a result of considering the above organisations' efforts in international action against cyber crime and cyberterrorism, the major themes of these organisations' efforts become apparent. The most significant treaties in this case are the Convention on Cybercrime and the Convention on the Prevention of Terrorism. The general purpose of the Convention on Cybercrime is laid down in the Preamble which is to deter crimes against the confidentiality, integrity, and availability of information systems and the misuse of such systems.

---

[134] Ministry of Justice, 8 April 2008, Letter No. 12-6/4620, *Re:* Judicial Cooperation between Estonia and Russian Federation.

The purpose of the Protocol is to supplement the provisions of the Convention on Cybercrime on the criminalisation of acts of a racist and xenophobic nature committed through information systems.[135] Although the Convention on Cybercrime and the Convention on the Prevention of Terrorism are categorised as regional efforts in combating cyberterrorism, they have prominent roles in this area and a number of countries which are located outside of their regions have ratified and became members of these Conventions.

Some of the organisations above promote security and try to prevent and remove cyber-related crime at the international level. Typical actions in this field have been taken by the UN and Interpol. Early in 1981, Interpol surveyed the criminal laws of member states so as to explore defects in the existing legislation, and made efforts to harmonise the laws. Today, Interpol's African Working Party on ITC Projects is trying to persuade the African states to sign and ratify the Convention on Cybercrime.[136] Besides this, the EU Framework Decision of 2002 specifically granted member states the responsibility of criminalising the offences of illegal access to and illegal interference with information systems.[137]

These conventions attempt to prevent cyber attacks by harmonising their member states' laws. The harmonisation of these laws must also include a substantive national criminal law provision in all countries that covers various terrorist acts for the prosecution of cyberterrorism. This situation is impossible without the effort of international and regional bodies. The UN's various conventions against terrorism and terrorism-related acts obligate states to enact substantive criminal law provisions.

The existing international conventions and other legislations regarding the harmonisation of national laws and international cooperation are applicable to the misuse of the internet for terrorist purposes. The only issue is the existence of terrorist-specific gaps in

[135] Protocol of the Convention on Cybercrime, Article 1.

[136] L. Xingan, 'International actions against cybercrime: Networking legal systems in the networked crime scene'.

[137] L. Xingan, 'International actions against cybercrime: Networking legal systems in the networked crime scene'.

computer-specific conventions and computer-specific gaps in terrorist-specific conventions. Although both of these conventions are applicable together, the existing general gap is not specific to the use of the internet for terrorist purposes in computer-specific and terrorist-specific instruments.

According to law, when a foreigner attacks a state, the state's jurisdiction cannot be applied against that foreigner. The state has to rely instead on the existence of bilateral and multilateral extradition treaties and legal agreements.[138] Considering that neither a harmonised regulation on enactments of countries nor any imperative binding treaty between countries on cyberterrorism exists, these countries use instruments of international law, such as the UN Charter, the Convention on Cybercrime, and protocols of NATO and Interpol, in case of a cyber attack or cyberterrorism incident. The insufficient number of parties is a common problem for all international instruments in fighting cyberterrorism such as the Cybercrime Convention and the Convention on the Prevention of Terrorism.

The signing, ratification, and implementation of these two conventions should be supported. Any additional courses of action undertaken in this context should not hinder or distract from these processes. CODEXTER has invited the Committee of Ministers to solve this problem and encourage states to sign, ratify, and implement the relevant conventions. The Cybercrime Convention and Convention on Prevention on Terrorism should be evaluated to assess their abilities in covering technical advances. This evaluation is a normal process that frequently leads to revision and updates, particularly when dealing with high risks issues such as cyberterrorism. Terrorist attacks against computer systems should be sanctioned by countries' domestic laws on data and system interference. Effective, proportionate, and dissuasive sanctions must be left to the national legislature to sentence aggravated offences of data interference or infrastructure offences.

---

[138] D. M. Creekman, 'A helpless America? An examination of the legal options available to the United States in response to varying types of Cyber-attacks from China', *The American University International Law Review* (2003) 17(3), p. 646.

Under the current situation, ensuring the effective implementation of the Cybercrime Convention and the Convention on the Prevention of Terrorism should be the main focus. New negotiations might jeopardise the increasing influence of these conventions on the global campaign against cyber crime and terrorism. If national legislators establish appropriate sanctions for cyber terrorists according to the guidelines of the Cybercrime Convention, the convention will be implemented effectively. The CoE's Convention on the Prevention of Terrorism would effectively focus on curtailing the widespread dissemination of illegal terrorist content on the internet.[139]

In conclusion, all international, regional, and national level organisations' major aim is to improve their security by harmonisation of legislation, coordination, and cooperation in law enforcement and utilisation of direct and indirect anti-cyberterrorism actions. However, despite efforts at an international level, they have also made attempts to promote security at the domestic level, as APEC encourages its member states to do, to promote cyber security and to tackle the threats of cyber terrorists. The Shanghai Declaration of 2002 supported measures to fight against misuse of information. As discussed above, various organisations have coordinated and cooperated in law enforcement.

2.6 CONCLUSION

Combating cyberterrorism by using international legal efforts has two aspects. The first one is based on international organisations' activities and strategies to counter cyberterrorism and the subsequent aspect is the measures that have been taken by countries based on regulations of international organisations. States must adopt the necessary measures regulated by international organisations. International cooperation counters cyberterrorism, leading to an effective and preventive ability to prosecute online crime and cyberterrorism offences that

[139] The Committee of Ministers of the Council of Europe took note of the opinion of the CODEXTER on cyber terrorism and the use of the Internet for terrorist purposes at its 1019th meeting on February 27 and 28, 2008.

cross the international border. If countries want to be protected from cyber attacks and cyberterrorism, they have to cooperate, coordinate, and harmonise in international legal efforts. To do so such countries' coordination and harmonisation efforts are divided into two main areas. First is the legislation process combating cyberterrorism that requires countries to legislate according to elements that are drawn from the international harmonisation effort. The second step is global investigation and prosecution of cyberterrorism and the adoption of procedural laws to prosecute criminal offences. This step should apply to the collection of evidence in electronic form of criminal offences, an enumerated list of conduct to criminalise and finalised by countries in order to reach to certain amount of immunity in cyber space.(the least of offences that must criminalised, enumerated.) The various measures illustrated in this chapter indicate the need for laws to be harmonised to prevent transnational criminals from exploiting jurisdictional and legal loopholes among countries, providing fewer opportunities for them.

CHAPTER III

APPLICATION OF LEGAL PROVISIONS IN THE CASE OF CYBERTERRORISM

3.1 INTRODUCTION

The internet provides a global platform from which cyber terrorists can pose direct and indirect threats to national security. Moreover, the internet provides a distinctive target that is more immense than any physical target without the need for perpetrators to sacrifice their lives.[1] The internet has had a vast impact on global progress, and countries are becoming information societies. These changes have rendered critical information vulnerable.[2] The process of globalisation has also driven efforts to secure cyber space through an immense legal global response.

---

[1] K. A. Gable, 'Cyber-apocalypse now: Securing the internet against cyber terrorism and using universal jurisdiction as a deterrent', *Vanderbilt Journal of Transnational Law* (2010) 47, p. 66.

[2] *Countering The Use Of The Internet For Terrorist Purposes-Legal And Technical Aspects* United Nation Counter Terrorism Implementation Task Force (CTITF), CTITF Publication Series, United Nation, 2011, p. 3.

Consequently, countries have been testing a variety of methods to legislate efficient systems of cyber crime deterrence.[3] Terrorist use of cyber space to launch their attacks has created a problem that may require countries to either use existing legislations or create new laws. These laws enable the government and internet service providers to monitor and prevent this kind of conduct in the virtual world. In most countries, no law is specifically devoted to combating cyberterrorism. These countries address the issue of cyberterrorism using current legislations on conventional terrorism and cyber crime. If a computer attack is attributable only to a private citizen and no connection or sponsorship by any state is determined, the attack is considered a criminal matter regardless of the specific target or the resulting damage.[4]

3.1.1 Objective of the Chapter

This chapter aims to study and analyse the available legal provisions of related countries that may be applicable to cyberterrorism. All countries have three options in applying legal provisions to cyberterrorism cases: they can apply cyber crime legislations to internet-based terrorist acts; apply terrorism related legislation to cyber terrorist attacks; and to enact new legislation regarding terrorists' use of the internet. This chapter also provides a broad discussion that delineates ancillary cyber terrorist activities. The objectives include:

(1) To identify the available legislations relating to cyberterrorism in the US, the UK, and Malaysia; and
(2) To analyse the applications of the relevant provisions to the case of cyberterrorism.

---

[3] J. S. Schjolberg *et al.*, *Wanted: A United Nation Cyber Space Treaty' Global Cyber Deterrence*, East West Institute, United States, 2010, pp. 16–18.

[4] D. M. Creekman, 'A helpless America?', An examination of the legal options available to the United States in response to varying types of cyber-attacks from china', *American University International Law Review* (2002) 17(3), p. 661.

3.2 THE ELEMENTS OF CRIME FOR PROSECUTING VIRTUAL CRIME

For an individual to be prosecuted for a 'real-world' crime, four elements of a criminal activity must be satisfied beyond reasonable doubt:

(1) The prohibited conduct (the *actus reus*);
(2) The mental state (the *mens rea*);
(3) The specified circumstances; and
(4) The proscribed result or harm.[5]

For an individual to be prosecuted for a "virtual-world" crime however, the virtual crime must involve conduct that would make up a crime in the real world. Thus, if there are no real world consequences, then the virtual "crime" cannot be prosecuted.[6]

Therefore, until a criminal legislation is modified to deal with virtual crimes, crimes in the virtual world have to satisfy the same elements above for crimes in the real world. Therefore, first, the *actus reus* must be present when the perpetrator commits an illegal act. This act may occur fully in the real world, fully in the virtual world, or partially in both worlds. Thus, the illegality of the act is important, not where the act occurs.[7] Second, the *mens rea* (located in the real world) is made out when the perpetrator knew his conduct was in fact illegal. Third, the specified circumstances are made out when the criminal engages in that specific conduct although he is not legally entitled to do so.

The most difficult requirement to satisfy is the "proscribed result or harm" element. This is because an individual will suffer harm only if the virtual crime has a real-world impact. Real-world harm may

[5] S. W. Brenner, 'Is there such a thing as "virtual crime"?', *Criminal Law Review Journal* (2001) 4, p. 33. Available at: http://boalt.org/CCLR/v4/v4brenner.htm (3 July 2012).
[6] S. W. Brenner, 'Fantasy crime: the role of criminal law in virtual worlds', *Vanderbilt Journal of Entertainment & Technology Law* (2008) 11(1), p. 53–60.
[7] S. W. Brenner, 'Is there such a thing as "virtual crime"?', p. 33.

include both financial harm (e.g. transfer or loss of money in the real world) or emotional harm (e.g. conduct resulting in mental pain and suffering for an individual in the real world).

The problem here is that some cyber crimes may have consequences in the real world but these consequences do not satisfy the 'harm' element required to constitute a crime under current criminal codes. For example, for such virtual terrorist activities such as communication, recruitment, or hoarding virtual weapons, it would be difficult to prove that these result in harm in the real world. However, if any conduct relates to these acts, for e.g. preparation to commit these acts, actually leads to real world terrorist activity, or if the virtual terrorist activity ends up being implemented in the real world, and this can be proven, then the perpetrators may be prosecuted in the real world because real world harm can be established. Then, if real world harm can be proven, any virtual world crime could also be prosecuted. However, it must take into account that certain kinds of harm are not direct harm, but the proscribed harm may indirectly have effects or consequences for the life, health, and welfare of persons. This is the exact thing that happens in cyber crimes. Cyberterrorism as an example of these kinds of crime may cause serious threat to health, life, and well-being, when it comes in to circumstances of shutting down hospitals and healthcare facilities.[8]

This still leaves the problem of virtual terrorist conduct that only affects users in the virtual world, such as terrorist communication recruitment. However, although these offenders may not be prosecuted in the real world, they can be punishing by the virtual world itself, for example by either having the host deactivate the users' accounts or by banishing them from the virtual world altogether. In fact, such attacks must originate from a government not a private citizen in order be to labelled as cyberterrorism.

---

[8] M. Cherif Bassiouni, 'Crimes against humanity: The case for a specialized convention' *Washington University Global Studies Law Review* (2010) 9(4), p. 590.

3.3 OVERVIEW OF TERRORISM AND CYBERTERRORISM LEGISLATIONS FOR RESPONDING TO CYBERTERRORISM

3.3.1 The US

Over the last few years, the US has been the target of terrorist attacks on their network systems. Terrorists target network systems since this can have large-scale impact on nations.[9] It is also crucial to understand the interrelationship between physical and cyber security in the current technological environment. Coordinated attacks on multiple regions could achieve a national effect. Nowadays, terrorists increasingly use the power of modern communications technology for planning, recruiting, propaganda purposes, enhancing communications, command and control, fund raising and funds transfer, information gathering, and the like.

The Federal Bureau of Investigation (FBI) has announced that the cyberterrorism threat to the US is expanding rapidly, as the number of actors with the ability to utilise computers for illegal, harmful, and possibly devastating purposes is on the rise. Terrorist groups also develop or hire hackers, particularly for the purpose of complementing large physical attacks with cyber attacks.[10]

Therefore, the FBI has established the following initiatives to combat cyberterrorism: Cyber Task Forces, Public/Private Alliances, International Cyber Investigative Support, Mobile Cyber Assistance Teams, Cyber Action Teams, Cyber Investigators Training, a Cyber Intelligence Center, and Cyber Tactical Analytical Case Support.

---

[9] Testimony Of Keith Lourdeau, Deputy Assistant Director, Cyber Division, FBI, Before the Senate Judiciary Subcommittee on Terrorism, Technology, and Homeland Security, February 24, 2004. Available at: http://www.fbi.gov/news/testimony/hearing-on-cyber-terrorism (26 Jun 2013).

[10] Testimony Of Keith Lourdeau, Deputy Assistant Director, Cyber Division, FBI, Before the Senate Judiciary Subcommittee on Terrorism, Technology, and Homeland Security, February 24, 2004. Available at: http://www.fbi.gov/news/testimony/hearing-on-cyber-terrorism (26 Jun 2013).

These programmes provide a strategic framework and programme management tool for all FBI computer intrusion investigations. The Cyber International Investigative programme creates the ability to conduct international cyber investigative efforts through coordination with the FBI Headquarters Office of International Operations, Legal Attache offices, and foreign law enforcement agencies.

The Cyber Specialised Training Program coordinates with the Engineering Research Facility, Laboratory Division, Training Division, National White Collar Crime Center, private industry, academia, and others to deliver training to FBI cyber squads, task forces, international law enforcement officers, and others. The Director of the FBI has established new priorities protecting the US from terrorist attacks and protecting the US against cyber-based attacks and high-technology crimes. The FBI Cyber Division's designated Criminal Computer Intrusion Unit addresses computer intrusions.

Over the past two decades, with the increasing number of cyber crimes, the US government has attempted to pass new legislation and rework current legislation to combat these specific crimes. This has enabled prosecutors to "swiftly trace a cyber-attack back to its source and appropriately prosecute" without the need to deconstruct and rework the entire US Code. They are authorised to investigate computer crimes like bank fraud, counterfeit checks, credit card fraud, virus and worm proliferation, cyber attacks and computer system intrusions, and identity theft, all of which could be used to support terrorist funding or activities.

The US provides a cyber terrorist regime in Title 18 of the US Code, which has been amended by the USA Patriot Act of 2001 and the Cyber Security Enhancement Act of 2002 (CSEA). In fact, it reveals a measured approach to criminalising acts of cyberterrorism. The subsequent CSEA of 2002, which was signed into law as part of the Homeland Security package, improves upon the Computer Fraud and Abuse Act of 1986 (CFAA) by requiring stiffer penalties for computer crimes than were previously imposed.[11] Most of the

---

[11] T. Mythri Raghavan, 'In fear of cyber terrorism: An analysis of the congressional response', *Journal of Law, Technology & Policy* (2003) 1, p. 298.

Patriot Act 2001 is the amendment of the federal crimes of terrorism in Title 18 — the CFAA 1984. The act of cyberterrorism would fall under Section 1030 of Title 18 that is amended by the "deterrence and prevention of cyberterrorism" provision in Section 814 of the Patriot Act 2001. Section 1030 defines a "federal crime of terrorism" under Section 2332b (g)(5) for the purposes of investigating authorities where such an act is "calculated to influence or affect the conduct of government by intimidation or coercion, or to retaliate against government conduct".[12] Therefore, it is necessary to consider the CFAA and Patriot Act 2001 and Homeland Security Act 2002 separately (although they affect and complement each other in some sections and circumstances) in order to examine the facts that exist with respect to computer crime and terrorist crime in the US.

A. CFAA 1984

Before the enactment of several computer crime regulations, the US government had relied on its federal criminal law, in the same way as in many other countries, to protect against cyber threats. The first federal law which held people responsible for electronic crimes were the wiretapping statutes. For example, while email was considered to be "wire communication" under this statute, courts interpreted the law to require acquisition of any wire communication while it was being sent. In the case of email, however, this is often not possible, and consequently, the federal wiretapping statute provided insufficient remedies for such fraud. After that, in 1984, the CFAA was first enacted and it addressed computer crimes in federal statutes.[13]

The first attempt to address computer related crimes by Congress was the CFAA 1984. It has expanded in the last two decades to cover new unlawful acts. The types of computer attacks and the methods of these wrongs have expanded in the last two decades and the CFAA has been amended to cover these new unlawful acts.

---

[12] 18 U.S.C. 2332b(g)(5)(A).

[13] T. Mythri Raghavan, 'In fear of cyber terrorism: An analysis of the congressional response', p. 298.

B. Patriot Act 2001

The most prominent Act after the September 11 attacks is the Patriot Act 2001, which amends different sections of the US Code in various ways in regards to cyber crime and cyberterrorism.[14] The Patriot Act 2001 was passed by a majority vote in both Houses. It was passed 2 months after the September 11 attacks and included several amendments to existing federal laws.[15] Terrorism is addressed in the US Federal Criminal Code, Chapter 113B of Part 1 of Title 18. After 11 September 2001, the US Congress passed the Patriot Act 2001, which amends parts of Section 2331 of Title 18 and redefined terrorism. The Patriot Act 2001 added a definition for 'domestic terrorism', as well as adding 'mass destruction' as a method of conducting terrorism. 'International terrorism' is defined in Section 2331(1).

Under the Patriot Act (2001) terrorism is categorised as 'domestic terrorism' and 'international terrorism'. The former relates to activities that

> (A) are dangerous to human life that are a violation of the criminal laws of the United States or of any State; (B) appear to be intended (i) to intimidate or coerce a civilian population; (ii) to influence the policy of a government by intimidation or coercion; or (iii) to affect the conduct of a government by mass destruction, assassination, or kidnapping; and (C) occur primarily within the territorial jurisdiction of the United States.[16]

International terrorism under the act relates to:

> (A) violent acts or acts dangerous to human life that are a violation of the criminal laws of the United States or of any State, or that would be a criminal violation if committed within the jurisdiction of the United States or of any State; (B) appear to be intended (i) to intimidate or coerce a civilian population; (ii) to influence the policy of a

[14] T. Mythri Raghavan, 'In fear of cyber terrorism: An analysis of the congressional response' *Journal of Law, Technology & Policy* (2003) 1, p. 301.
[15] T. Mythri Raghavan, 'In fear of cyber terrorism: An analysis of the congressional response', p. 301.
[16] Section 18(I)(113b) United States Code (Act 2331).

government by intimidation or coercion; or (iii) to affect the conduct of a government by mass destruction, assassination, or kidnapping; and (C) occur primarily outside the territorial jurisdiction of the United States, or transcend national boundaries in terms of the means by which they are accomplished, the persons they appear intended to intimidate or coerce, or the locale in which their perpetrators operate or seek asylum.[17]

C. CSEA 2002

The subsequent act following the Patriot Act 2001 is CSEA 2002, which was signed into law as part of the Homeland Security package, also improves upon the CFAA 1984 by requiring stiffer penalties for computer crimes than were previously imposed. President George W. Bush signed the Cyber Security Research and Development Act on November 27, 2002, which promotes research and development in the relatively unexplored and underfunded area of cyber security. In 2003, several other bills were introduced that also addressed cyber security issues. In fact, the government had concerns prior to the September 11 attacks, but this incident was a catalyst that accelerated cyber security concerns.[18]

The Federal CFAA protects federal computers, bank computers, and computers connected to the internet. It outlaws conduct that victimises computer systems. Being new versions of the Act, they may not be comprehensive enough provisions but they fill cracks and gaps in the other criminal protection laws. In particular, Sections 1030(a)(2), (3), (4), (5), and (7) are related to attacks against government computers.[19]

D. ECPA

The ECPA 1986 is a law enacted by the US Congress in 1986 that sets provisions on what privacy rights people have when they use telephones,

---

[17] Section (803) USA Patriot Act (2001) (Act 2339).

[18] T. Mythri Raghavan, 'In fear of cyber terrorism: An analysis of the congressional response', *Journal of Law, Technology & Policy* (2003) 1, p. 301.

[19] C. Doyle, Cyber-crime: An Overview of the Federal Computer Fraud and Abuse Statute and Related Federal Criminal Law, 97–1025, CRS Report for Congress, 2010, p. 5.

computers, cell phones or other means of electronic transmission of communication, like faxes or texting. It was the amendment of the Omnibus Crime Control and Safe Streets Act of 1968 which attempted to balance between the privacy rights of individuals and the legitimate needs of law enforcement.[20] It sets out the provisions for access, use, disclosure, interception, and privacy protections of electronic communications. The law was enacted in 1986 to include electronic communications and privacy within the originally intended protections of the Federal Wiretap Act 1968 and covers various forms of wire and electronic communications.[21] When this Act was enacted, the provisions of the ECPA did not include some of the newer forms of communication developed since then.

3.3.2 The UK

The present UK government is in the best position in the world in the area of utilising the advancements of ICT. It empowers its companies, government ministries, and business transactions with ICT in a way that was unavailable to its predecessors and which generates a new era of an emerging *e*-society. In addition, the UK is the second largest software consumer in the world. Also, 80% of the UK's population has access to broadband.[22]

Electronic cyber attacks were identified by the UK with this notion: they usually involve using computers to gain unauthorised access to data or control of software or hardware and denial of service (DoS). These methods are widely available and many can be downloaded from the internet. Sophisticated attacks are unlikely to be detected by routine security measures such as firewalls and internet detection systems.[23]

---

[20] S. K. Rahavy, 'The federal wiretap act: The permissible scope of eavesdropping in the family home', *Journal of High Technology Law* (2003) 88, pp. 87–88.

[21] M. Sherman, *Cyber-crime and Cyber Terrorism*, Federal Judicial Centre, 2002, p. 1.

[22] P. Wilkinson, *Homeland Security in the UK: Future Preparedness for Terrorist Attack since 9/1*, 1st Edn, Routledge Publishing, New York, 2007, pp. 50–55.

[23] P. Wilkinson, *Homeland Security in the UK: Future Preparedness for Terrorist Attack since 9/11*, p. 296.

Cyber attacks, terrorism, and cyberterrorism and the fear that they generate have been the greatest threats to British security.[24] In the past decade, cyber attacks have been committed by normal people and their targets have not been vital ones. The perpetrators were motivated by money to commit these crimes. In reflection, today the focus of cyber attackers has shifted from these "normal" things to critical national infrastructures such as power grids, telecommunications, banking networks, and water and electricity networks.

As mentioned above, cyberterrorism is considered a type of cyber crime and the legal measures for prosecuting and investigating cyber crime are applied to cyberterrorism as well. The UK's strategy to tackle terrorism is to criminalise specific actions rather than the medium through which the actions are committed. Therefore, terrorist related actions are unlawful in the UK, but it is not limited to those that take place in cyber space. Put in another way, the UK has not planned new legislation to specifically counter terrorists' misuse of cyber space. Most detection and monitoring in the UK is carried out by organisations and competent authorities.

In the UK, people who endanger lives through the manipulation of public computer systems are considered under the anti-terrorism law, which was enacted in February 2001.[25] The UK attempted to introduce anti-terrorism regulation that would have obliged all local and national government agencies to gain access without warrant to communications traffic data.[26] In order to respond to terrorism and cyberterrorism, terrorism related Acts and the Computer Crime Act are used. The terrorism related Acts in the UK include the Terrorism Act 2000, the Anti-terrorism, Crime and Security Act 2001 and the Bill for the Prevention of Terrorism Act 2005 (which came into force as the Terrorism Act 2006).

In the UK, there are two important Acts that give the police exceptional powers to deal with extraordinary situations: the Terrorism

---

[24] Ministry of Defence website, https://www.gov.uk/government/organisations/ministry-of-defence (12 Jun 2013).

[25] Available at: http://www.crime-research.org/latestnews/ (17 Jan 2011).

[26] B. M. Jenkins, *The New Age of Terrorism*, 1st Edn, RAND Publication, Santa Monica, California, 2006, pp. 126–128.

Act 2000 and the Anti-terrorism, Crime and Security Act 2001. The former Act was enacted as a response to the September 11 attacks. It was enacted to amend the Terrorism Act 2000, to make further provisions regarding terrorism and security, and to extend the criminal law and powers for preventing crime and enforcing that law, on 14 December 2001, following the September 11 attacks. In fact, this Act allows the police to detain a foreign suspect without charge if the individual cannot be deported for other legal reasons.[27]

In UK legislation, the first Act that addressed computer crime was the Computer Misuse Act 1990. It became law on 29 August 1990. The Act originated directly from the "Law Commission Report on Computer Misuse" that was published in October 1989. This Act announced three new categories of offences that foresee cyber terrorists implementing their actions by using these methods: unauthorised access to computer material, unauthorised access with intent to commit a further offence, and unauthorised acts with intent to impair. However, when new kinds of attacks began to emerge with enhanced technologies, the perceived utility of the Act decreased and the Act was found unsuitable to deal with the range of problems that were produced by the advent of the internet. Finally, such problems led to a series of amendments introduced by the Police and Justice Act 2006. Section 3 of this Act was broadened to refer to impairment of the system, and Section 3A defined a new offence in relation to making, supplying, or obtaining articles for use in offences under Sections 1 or 3.

Cyber terrorist organisations utilise their cyber attack tools by engaging hackers in order to acquire their skills in launching their attacks. However, the institutional attack that organised cyber terrorists launch is completely different from that of the recreational hacker who carries out this behaviour "as a proof of accomplishment and acquisition of skill set in which thrill of the challenge was the only goal". It must be noted that a hacker could be a cyber terrorist if he uses a cyber attack tool to hack the targeted goal with a specific

---

[27] Available at: http://www.new.bbc.co.uk/2/hi/uk_news/3197394.stm (18 Jan 2011).

intention, but it does not mean that all hackers are cyber terrorists. Once the hacker penetrates a computer system, he could do several different things, such as read or copy information, erase or modify information or programs stored in the computer systems, download programs or data, or he could even add something. Given the fast-growing pace of internet technology, a great deal of concern has been raised that the Computer Misuse Act has failed to keep pace with development and that it remains ineffective against new forms of cyber crime.

A. Terrorism Act 2000

The Terrorism Act of the UK, meanwhile, dates back to 1983. The advanced version of terrorism in Section 20(1) of the Prevention of Terrorism Act 1989 states that: "terrorism means the use of violence for political ends and includes any use of violence for the purpose of putting the public or any Section of the public in fear". In 1983, an official verdict announced that Section 20 "has not given rise to any difficulties" and that it did not require amendment. Then, in 1996, Lord Liyon said that it was too narrow and could not catch every single issue of terrorism. Therefore, Section 1 of the Terrorism Act 2000 expanded both the forbidden activities and feared consequences. However, this definition was an administrative definition which served for the purpose of the Act. It did not have legal terms and did not define criminal offences. It did not create the criminal offence of terrorism, since generally, terrorism offences include activities involving the commission of ordinary criminal acts.[28]

Following 11 September 2001, the Anti-Terrorism, Crime and Security Act 2001 was the UK's initial response. This was because; the previous Act (Terrorism Act 2000) was not able to deal with the threat of international terrorism and religious extremism. The Anti-Terrorism, Crime and Security Act 2001 was introduced into parliament on 19 November 2001, and came into force on 14 December 2001.

[28] A. Aust, 'Counter Terrorism — A new approach, the international convention for the suppression of the financing of terrorism', *Max Planck Yearbook of United Nations Law* (2001) 5, p. 290.

It amended the terrorism act 2000 with the aim of making further provision about terrorism and security, providing freezing for assets, and providing the retention of communication data. The Anti-Terrorism, Crime and Security Act 2001 tightened the legislation on terrorist property. The Prevention of Terrorism Act 2005 was established to address the unfair control order scheme. It redressed it by introducing a new form of control order which is applicable to UK citizens as well as foreign nationals. The previous one was only applicable to non-UK nationals.[29]

The later Terrorism Act 2006 criminalised training and encouragement of terrorism, aiding and abetting terrorism, and dissemination of terrorism and propaganda that could cause others to commit terrorist acts. Preparation, attending terrorist training camps, or purchasing equipment intended to commit a terrorist act were also criminalised according to new Act.[30] It addressed terrorism not only in the UK, but also anywhere in the world.

B. Computer Misuse Act 1990

The Computer Misuse Act was enacted after the case of 1984–1985 *R v. Gold*, which was appealed in 1988. This case ultimately caused the enactment of the Computer Misuse Act 1990. The case occurred when an employee of Prestel at a tradeshow entered his username and password from his home with one of his colleagues whose name was Schifreen. They used this information from their home computer to access the British Telecom Prestel system, and then they entered into the private message box of Prince Philip. Prestel became aware of this access, trapped the two men and charged them with fraud and forgery. The two men were convicted and fined, but they appealed their case. They appealed their conviction based on the claim that they were not using the data for either personal or illegal gain. They did not gain any material benefit by spying on someone else's system.

---

[29] M. Charvat, 'A study of UK Anti-terror law, legal aspect of combating terrorism' *In Centre of Excellence Defence Against Terrorism*. IOS press, Ankara, Turkey, 2008, pp. 109–110.

[30] M. Charvat, *Legal Aspect of Combating Terrorism*, pp. 109–110.

They argued that they did not deserve to be charged under a specific law and it could not apply to them. Finally, the House of Lords acquitted them, but the successful appeal led to the passing of an Act to forbid such behaviour in future. Consequently, the Computer Misuse Act passed into law in 1990.

The Computer Misuse Act is aimed at securing computer material against unauthorised access or modification and related actions. It is meant primarily to manage the problem of computer hackings especially with the advent of the internet. Among the provisions, the act makes it an offence for anyone to access computers they have no authority to use. It does not specify any target that the unauthorised user may have and makes it unlawful, for example, to run port scanners in an attempt to find insecure computers. This applies to English law. Under Scottish law, computer intrusion falls under the ambit of the common law related to deception. Three criminal offences are listed under the act, namely, unauthorised access to computer material, access aimed at committing or facilitating the commission of further offences, and modification of computer material that is unauthorised.

3.3.3 Malaysia

The Malaysian government realises that the world has changed and that the threats influencing critical infrastructure have markedly increased. In particular, the traditional threats have evolved into new cyber threats. Therefore, the Malaysian government has adopted an integrated approach and policies to protect its national infrastructure from cyber threats. These policies have moved towards controlling vital assets which comprise the networked information systems of critical sectors, including banking and finance, information and communication, energy, transportation, water, health services, government, emergency services, and food and agriculture. These policies, to achieve proper protection of vital assets are being implemented in various ways. The government promotes cooperation between the public and private sectors. There are also Malaysia cyber laws to deal with cyber security threats and Malaysia also harmonises these with international laws and treaties and moves towards strengthening enforcement in a standard form. The government has also established and strengthened a national computer emergency response team.[31]

Another governmental policy is to encourage the gathering of critical national infrastructure information (CNII) to monitor cyber security events, to design vulnerability assessment programmes, and to establish mechanisms for cyber security knowledge dissemination at the national level. Since it is also very important for every organisation to have coherent protection mechanisms, the Malaysia government has mandated CNII organisations to implement Information Security Management System (ISMS) based on ISO27001 standards. The Organisations have been given a 3 year period (2011–2013) to implement the system. However, the level of compliance even after the due date has not been satisfactory. The implementation approach is a 9-year programme which is divided into three levels. The first level is to achieve immediate concern and it lasts 1 year. This phase creates a platform for security mechanisms, raises awareness of cyber security and covers the gaps that exist in cyber security of the CNII to address fundamental vulnerabilities.[32]

Phase 2 builds the infrastructure, sets up system processes, and creates capability amongst researchers and information security experts. This phase lasts 3 years. The last step is to develop self-reliance. This phase tries to develop technology, evaluate mechanisms and create a culture of cyber security, and lasts for 5 years.

Malaysia has established Cyber Security Malaysia to achieve these policies. It began in 1997 as the Malaysian Computer Emergency Response Team then it became the national ICT Security and Emergency Centre in 2007 and then changed its name to Cyber Security Malaysia. The body provides support to enforcement agencies and victims in ensuring that justice will prevail regardless of the "space" where a particular crime is committed. It also created a help centre named Cyber999 Help Centre that provides services for

[31] K. Mitnick, *E Security the First Line of Digital Defense Begins with Knowledge*, Vol. 26, Malaysia, 2011, pp. 5–7. Available at: https://www.cybersecurity.my/data/content_files/12/852.pdf.

[32] The national Cyber Security Policy. Available at: http://cnii.cybersecurity.my/main/ncsp/tncsp.html (9 March 2014).

internet users such as digital forensic and data recovery services, a cyber security strategy policy, and legal responses.[33]

A. Penal code of Malaysia

The amendment to the Penal Code was created to enable Malaysia to accede to the International Convention for the Suppression of the Financing of Terrorism adopted by the General Assembly of the UN on 9 December 1999 (ICFST). It was amended to comply with Security Council Resolution 1373 (2001) on 28 September 2001 (UNSR). Malaysia ratified the Charter of the UN and has been a member of the UN since 1945. Malaysia established the Human Rights Commission of Malaysia in 1999 by the Human Rights Commission of Malaysia Act. Suara Rakyat Malaysia (SUARAM) allocates all efforts to combat terrorism by calling for the legislating of criminal law and laws strengthening security. It tries to propagate human rights principles and the norms of civil society which originate from international conventions and treaties.[34]

The International Convention for the Suppression of the Financing of Terrorism (ICSFT) offers wide powers to states to take the proper measures, which are adopted by international law and standards and international human rights. Furthermore, it broadens the criminal jurisdiction over terrorism offences according to the standards of international law. As discussed above, amendments to the criminal sections of the Malaysian Penal Code were made to conform to the standards of international organisations.[35]

The Malaysian Penal Code is the primary legislation consisting of 511 sections encompassing a broad range of criminal offences punishable in Malaysian courts of law. The Criminal Procedure Code (CPC) was enacted to regulate the administration of criminal justice or procedures in Malaysia. Among others, the CPC provides the procedures through which an arrest, search, police investigations, and trials are conducted in the country.

---

[33] K. Mitnick, *E Security the First Line of Digital Defense Begins with Knowledge*, pp. 5–7.
[34] Available at: http://ejp.icj.org/img/cpc_terrorism_amendment.pdf (22 Feb 2011).
[35] Available at: http://ejp.icj.org/img/cpc_terrorism_amendment.pdf (22 Feb 2011).

B. The security offences Act

Created under Article 149 of Federal Constitution this act aims to prevent internal security issues of public order, acts of terrorism, sabotage, and espionage. Malaysia has always had a specific Act that deals with terrorism or any form of threat to national security, namely the Internal Security Act 1960 which was drafted in accordance with Article 149 of the Federal Constitution. However, the Internal Security Act 1960 Act has been repealed and the government has attempted to enact a new rule modelled on the Patriot Act of the US and the Anti-Terrorism Act in the UK. The Internal Security Act was replaced by the Security Offences Act 2012. It provides special measures relating to security offences. Under Section 4 (the key provision) of the Act, a police officer has special powers of arrest and detention over persons whom he has reason to believe to be involved in security offences.

C. Computer Crime Act 1997

The Computer Crime Act was enacted to deal exclusively with computer crime and computer related crime. It was drafted by a committee from the Malaysian Attorney General's chambers following the model of the Computer Crime Misuse Act of Malaysia. The bill of the Computer Crime Act remained secret until March 1997 in the Dewan Rakyat, Malaysia's House of Representatives. The opposition party protested against this secrecy and requested for public discussion prior to the parliamentary debates. By denouncing the bill, the opposition party voiced concerns about some penalties that they believed were unreasonable. At last, the Act was adopted in June 1997. Under Part Two of this Act, the offences of computer crime are set out such that cyberterrorism can be covered by this Act to some extent.

3.4 LEGAL RESPONSES ACCORDING TO TERRORISM STATUTES

This section is created to differentiate between ancillary cyber activities and cyberterrorism. Ancillary cyber activities means The facility which is provided by terrorist groups or individuals to conduct their activities. Although, these activities are not considered as cyberterrorism attacks, they can be used to support terrorist activities that lead to the launch of a cyber terrorist attack. It is important to consider them carefully as they are not the same as a pure cyberterrorism attack.

3.4.1 Ancillary Cyber Activities from the Perspective of Relevant Countries

Cyber terrorists utilise the internet in two ways. Terrorists recognise the benefits of cyber operations to launch their disruptive attacks; therefore, they exploit information technology in every part of their operations. They also use computers to support their political or military objectives. These activities cannot be described as "terrorising" activities *per se*, and legal action under specific laws should not be taken against them unless they are linked to a terrorist group. That is, legal action should only be taken to reduce the capability of terrorist organisations to terrorise by other methods.[36] However, the problem that confronts us is if these activities are not linked to a terrorist group, then there is the risk of infringing into political activism, which should not be viewed as terrorism.

As will be seen, almost all countries do not include ancillary terrorism and cyberterrorism activities in their terrorism acts. Although they consider cyberterrorism as part of terrorism, or a new method of terrorism, ancillary cyber activities relating to terrorist organisations are not considered as cyberterrorism. Rather, these activities consider how terrorists use the internet and the impact of such use as terrorism. Terrorists use the internet to facilitate their activities, inflict harm, and distribute information. They expand their options of attack rather than just replace the traditional way of operating the terrorism activities.[37]

---

[36] C. Walker, 'Cyber terrorism: Legal principle and law in the United Kingdom', *Penn State Law Review* (2006) 110, p. 640.

[37] D. E. Denning, Terror's Web: How the Internet is Transforming Terrorism, In: Y. Jewkes and M. Yar, (Eds.), (1st Edn.) *Handbook on Internet Crime*, Willan Publishing, United States, 2009, pp. 55–60.

These two identical notions should not be confused with each other. Although both of them are considered as cyber activities, we cannot place them in the same pot. Cyberterrorism differs from cyber activity and terrorist use of the internet. The use of the internet by terrorists merely helps terrorist to carry out cyber attacks, but that is not cyberterrorism activity. Terrorism has been transformed through cyber attacks. Terrorists obtain hacking tools and information from the internet and utilise them to launch their cyber attacks. These are specifically the actions that we call terrorist use of the net or ancillary cyber terrorist activity, because they support terrorist objectives, although these acts are not characterised as cyberterrorism. The internet has introduced a location from which hackers can inflict damage without engaging in violence.[38]

Experts believe that terrorists use the internet for the following objectives: research and communication, training, fundraising, media operations, radicalisation, and recruitment.[39] The internet provides a medium for terrorist groups that replaces and broadens their traditional social networks. Ancillary cyber activities are used to launch cyberterrorism but are not considered as acts of cyberterrorism. The related provisions on ancillary cyber activities are discussed in the next section. These activities focus on ancillary terrorism activities that can be applicable to cyberterrorism as previously said.

A. The US

Under paragraph 1030(a) of the US Code, the crimes of attempt, conspiracy and complicity are the same as other crimes and subject to the same penalty. Put another way, those who attempt to, or aid and abet the violation of another are subject to the same penalties as those who commit the substantive offence.[40]

---

[38] Y. Jewkes *et al.*, *Handbook on Internet Crime*, 2nd Edn, Routledge Publisher, United Kingdom, p. 122.

[39] CQ Researcher, *Issues in Terrorism and Homeland Security: Selections from CQ Researcher* 2nd Edn, SAGE Publication, United States of America, 2011, p. 130.

[40] Section 1030(b) United States Code.

It is worth noting that, although the support of terrorist activity in all regions and states is considered as terrorism and the terrorism definition encompasses the support of terrorist activity, ancillary terrorist activity via the internet and terrorist activity on the internet are not labelled as cyberterrorism. The cyberterrorism definition does not cover ancillary cyber activities. In fact, terrorist use of the internet furthers and secures their organisational goals and enables them to operate and grow with a greater degree of efficiency. They exploit modern tools in order to further the goals they seek.

They enhance their operations by employing digital technologies. Terrorists facilitate terrorism activity through the use of digital technology. A number of examples illustrate the appeal of this technology for terrorist groups to advance their particular agendas. Terrorist groups in Asia and Western Europe have exploited the web, such as the Neo-Nazi group, the Colombian ELN, the Zapatists, and the Hezbollah. Supporters of terrorists broadcast videos online and even add gruesome details for a horrifying effect. Notably, ancillary terrorist activities are considered as terrorist acts in most regions, but are not categorised under cyberterrorism. An attempt at launching an attack does not necessarily end up as a full-scale cyber attack (the main characteristic of cyberterrorism). Hence, a person committing such activity can only be convicted of preparing or aiding a terrorist offense or preparing for a terrorist offence.

The use of the internet for propaganda and disinformation is a popular with terrorists. For example, training videos with instructions on building explosive devices, preparing gunpowder, tips on money laundering, and achieving organisational needs have appeared on websites regularly accessed by militant Islamic groups.[41] These websites are also used to post recruitment videos. For example, in June 2005, al-Qaeda in Iraq, an al-Qaeda affiliated terrorist group led by Abu Musab al-Zarqawi, posted the second edition of its

[41] Technology and terror: *The modus operandi*, Andrew Bwcher, January 2005, Available at: http://www.pbs.org/wgbh/pages/frontline/shows/front/special/tech.html (3 Jul 2012).

recruitment magazine on the internet. It is also suspected that terrorist organisations that use computers for communication have moved beyond hierarchical organisational structures and now employ networked ones.[42]

The EU passed three laws on 28 November 2008 that apply to the following online terrorist activities: (1) public provocation to commit a terrorist offence, (2) recruitment for terrorism, and (3) training for terrorism. Unlike other countries — such as those in the European Union — who have enacted three new laws addressing issues of terrorist propaganda, recruitment, and training on the web, the US has not enacted any law to directly confront the threat of terrorist use of websites and which can be applied to online terrorist activity such as recruitment and training. Although the Patriot Act 2001 provides many useful tools for combating terrorism, it does not say anything about online terrorist activity and fails to address the problem presented by terrorist websites. It does not provide a specific strategy to combat it, even if the White House or Congress has the intent to come up with a comprehensive strategy. The National Strategy to Secure Cyberspace and the CSEA 2002 under the Homeland Security Act offer comprehensive plans to secure the US against cyber terrorist attacks only on critical infrastructures.

The US must recognise the growing problem of terrorist propaganda, recruitment, and training on the internet in order to establish a strategy to disrupt terrorist internet activities. The new laws of the EU can be taken as a sign that other countries are dealing with this problem as a real issue. Given that the US and the EU had similar laws on terrorism prior to the new laws[43] and both lack a unified strategy against cyber terrorists, the US should follow the EU's implied strategy of rejecting the material support statutes by introducing the new

---

[42] M. Stohl, 'Cyber terrorism: A clear and present danger, the sum of all fears, breaking point or patriot games?' *Crime Law and Social Change* (2007) 46, p. 232. www.springerlink.com/index/Y816 117WW6058JP7.pdf (3 Jul 2012).

[43] Prior to the introduction of these new statutes, both legal regimes begin their terrorism laws with a definition of the basic "*mens rea*" for terrorist offenses as "intending to intimidate or coerce a population or government". Each piece of legislation then goes on to spell out similar specific conduct that will constitute a terrorist offense.

legislation to address public provocation, recruitment, and training for terrorism.[44] The EU created a counter-terrorism strategic plan in 2005 to criminalise public provocation to commit a terrorist offence. Freedom of speech protections prohibit the US from enacting a similar law. Although both the US and the EU claim to uphold freedom of speech, the US is stricter in protecting this human right.[45]

Two sections of the US Code which relate to material support are Sections 2339A and 2339B, especially Section 2339B, which is ideal for cyber-related activities. Section 2339A was enacted in response to the 1993 bombing of the World Trade Centre. It prohibits individuals or groups from providing material support or resources "knowing or intending" that they will be used to commit terrorist crimes. It defines "material support" as including, but not being limited to, the donation of money or financial services, training, communications equipment, and religious, materials. Due to the fact that proving that an individual or a charity donated money with knowledge or intent that it would be used to carry out a terrorist attack could be very difficult, Congress enacted Section 2339B as part of legislation entitled the Anti-terrorism and Effective Death Penalty Act (AEDPA) in 1996.

Section 2339B requires merely that an individual knowingly provide material resources to a group designated as a foreign terrorist organisation. It creates a separate offence for supporting terrorism that does not need to include any specific attack or event in order to qualify for prosecution of the activity. Thus, the federal prosecutor can charge terrorists and supporters early in their conspiracies, because there is no need to link 2339B offences to specific events or attacks.[46]

---

[44] M. A. Healy, 'How the legal regimes of the European Union and the United States approach Islamic terrorist websites: A comparative analysis', *Tulane Law Review Association* (2009) 84, pp. 173–174.

[45] The freedom of expression is safeguarded in the United States through the first amendments to the U.S. Constitution.

[46] M. A. Healy, 'How the legal regimes of the European Union and the United States approach Islamic terrorist websites: A comparative analysis', p. 171.

However, only one person has been prosecuted under Sections 2339A or 2339B. Oussama Kassir was convicted on 12 May 2009 for operating a terrorist website that spread information on how to build bombs and make poisons, and for trying to establish a terrorist training camp in the US. He was charged with providing material support to terrorists. He developed and operated several extremist websites which promoted terrorism and disseminated terrorist manuals, such as 'The Mujahiden Explosives Handbook'. Another case was that of Sami Al-Hussayen, who was a graduate student at the University of Idaho and who was arrested by the FBI in 2003 for creating, maintaining, and using websites allegedly linked to and materially supporting terrorist organisations. Following a bank teller's report to the FBI of suspicious activity, an investigation of him was begun by the FBI in 2001. The FBI discovered that Al-Hussayen had six bank accounts in four different states, and that he was the webmaster of three Islamic websites, two of which claimed to be Islamic charities.[47]

The FBI's investigation found that Al-Hussayen's offences on the internet included:

> (1) designing, operating, and maintaining websites for the Islamic Assembly of North America (IANA) and the Al-Haramain Islamic Foundation (AHIF) — both charities suspected by the FBI of fund-raising for terrorist organisations and spreading extremist Islamic ideas on their websites; (2) moderating an e-mail group to which he made personal postings extolling the virtues of violent jihad and encouraging young Muslims to donate money to the cause, and in which he had Webmaster power to control which postings remained on the site and which should be deleted; (3) creating an online system for donating money to Hamas, a Palestinian organisation infamous for its suicide bombings and designated as a foreign terrorist organisation by the United States; and (4) providing two Saudi clerics with websites

---

[47] A. F. Williams, 'Prosecuting website development under the material support to terrorism statutes: Time to fix what's broken', *New York University Journal of Legislation and Public Policy* (2008) 11, pp. 376–377.

publishing fatwas 'justifying and encouraging violent jihad, including suicide attacks'.[48]

He was thus charged with two counts of violating Section 2339A (one for conspiracy to provide material support and one for actually providing it) and another count of violating Section 2339B (for conspiracy to provide material support to Hamas). Al-Hussayen claimed that creating and maintaining websites that "were merely used to publish speeches" did not meet the definition of providing material support to terrorists, because they did not "give something with the intent that it be used in preparation for or in carrying out a predicate offense".

In response, the government argued that "the actions alleged in the indictment are more than simply creating, maintaining, or using websites but also providing money, expert advice, and other material support and resources intended to recruit and raise funds for terrorist actions". Al-Hussayen further argued that the creation of websites and email groups in 'virtual space' did not represent "providing communications equipment" (which means giving someone a physical device to be used for communication).

The government responded by stating that the definition of communications equipment included computer servers, which Al-Hussayen paid for and used to locate his websites. The US District Court for the District of Idaho denied Al-Hussayen's motion to dismiss, and found that Al-Hussayen's arguments were questions reserved for the jury and that it was up to the government to prove the required elements of the offences at trial. The court held that the indictment alleged specific actions of Al-Hussayen and that he was not merely a passive party. In addition, the court held that Al-Hussayen's definition of 'communication equipment' was too narrow, as Congress had intended the definition of 'material support' in Section 2339A to be interpreted broadly. Therefore, the case went forward to trial.[49]

---

[48] A. F. Williams, 'Prosecuting website development under the material support to terrorism statutes: Time to fix what's broken', pp. 376–377.

[49] *United States v. Al-Hussayen* [2004], U.S. Dist. LEXIS 29793, pp. 1–3.

In 2004, a jury found Al-Hussayen not guilty of all three charges of material support. At the trial, Al-Hussayen argued that his actions on the internet were protected by freedom of association and freedom of speech. Whether this argument swayed the jury or not, one juror said in an interview that there was no 'hard evidence' that Al-Hussayen was a terrorist. Thus, it appears that the jury did not believe Al-Hussayen's recruitment websites constituted material support to terrorism under Sections 2339A and 2339B. It was submitted that the fact that neither statute mentions the words 'computer', 'internet' or 'websites' played a large role in the acquittal. Thus, at this moment in time, federal prosecutors have only convicted one person of materially supporting terrorism by operating terrorist websites. This happens in other jurisdictions as well. For example, in the UK, these ancillary activities are considered to be supportive to terrorism.

B. The UK

The Terrorism Act of 2000 of the UK deals with ancillary cyber activities relating to prescribed organisations in Section 12. Sections 58 and 103 also deal with these activities. Section 58 describes a "person who collects or records or possesses information of a kind likely to be useful to a person committing or preparing an act of terrorism". Hamaad Munshi was convicted as a terrorist in 2008 under this law for possessing materials likely to be used for terrorism. Munshi participated in an online British extremist group forum that discussed plans to fight overseas.[50] He shared terrorist videos and used a chat room to incite others to fight. The leaders of the group incited Munshi to fight. He was arrested at the age of 16. He downloaded details on how to make napalm and grenades, as revealed in the trial.[51] As such, a person who collects, records, or possesses information of a kind likely to be useful to a person committing or preparing an act of cyberterrorism shall be punishable under the provision on ancillary cyber terrorist activities.

---

[50] CQ Researcher, *Issues in Terrorism and Homeland Security: Selections from CQ Researcher*, p. 130.

[51] CQ Researcher, *Issues in Terrorism and Homeland Security: Selections from CQ Researcher*, p. 146.

The Act defines the term "record" as "photographic or electronic forms" as well as "writing and drawings" which are tangible, but does not cover mental knowledge. Nevertheless, in Section 58(3), it is stated that defendants must prove that they had a reasonable excuse for their actions or possession. This clause will apply in the event possession is proved and it is unnecessary to show that the information was obtained in breach of law, or to show how it was obtained. In *R v. Lorenc*,[52] the possession of army manuals was the basis for conviction.

However, it is submitted that "information" here could also apply to legitimate users such as academics, researchers and scholars. Further, Section 103 disputes Section 58 of the Northern Ireland (Emergency Provisions) Act 1973. The *actus reus* of this Section is made out by "publishing", "communicating", or "attempting to elicit", as well as "collecting or recording".[53]

In the case of *R v. McLaughlin*,[54] a radio buff was acquitted because he had reasonable excuse for possessing a list of Royal Ulster Constabulary radio frequencies.[55] Further, the presumption in subsection (4) states that if it can be proved that a piece of information: "(a) was on any premises at the same time as the accused, or (b) was on premises of which the accused was the occupier or which he habitually used otherwise than as a member of the public", then the court could assume that the accused possessed that information, unless the defendant can prove that he had no knowledge of its presence on that premises or that he had no control over that premises.[56]

Section 12 of the Terrorism Act 2000 deals with support of cyber activities of proscribed organisations. Furthermore, the offence of possession of items useful to terrorism is considered in Section 57 of the Terrorism Act 2000. This originated from the Emergency

---

[52] EWCA Crim 3404 (2005). Available at: http://legislationline.org/documents/action/popup/id/7040 (1 Sept 2013)

[53] Section (103) Terrorism Act 2000.

[54] 19 A.R, (1979), p. 368.

[55] C. Walker, 'Cyber terrorism: Legal principle and law in the United Kingdom', *Penn State Law Review* (2006) 110, p. 644.

[56] Section (103) Terrorism Act 2000.

Provisions Act 1991 of Northern Ireland.[57] It applies to one form of computer use by terrorist groups, to support their political and military objectives.

Regarding possession in public places, the Criminal Justice and Public Order Act 1994 extended this to Britain. According to Section 57(1), a person commits an offence if he were to "possess an article in circumstances which give rise to a reasonable suspicion that this possession if for a purpose connected with the commission, preparation or instigation of an act of terrorism". The penalties are the same as for Section 54. The Anti-Terrorism, Crime and Security Act of 2001 was introduced to the parliament on 19 November 2001 and came into force on 14 December 2001. This law amended the Terrorism Act of 2000 with the aim of making further provisions on terrorism and security, such as the freezing of assets and retention of communication data, to tighten legislation on terrorist properties.

After the September 11 attacks, the concept of terrorism changed, since previously the Terrorism Act 2000 had addressed domestic problems of terrorism. By the Anti-Terrorism, Crime and Security Act 2001 the concept of international law was incorporated in a new section, addressing funds or property that were outside British jurisdiction. By introducing a "freezing order" in this new Act, the government was authorised to freeze the assets of persons or organisations suspected of being involved in terrorism. This enabled the Treasury to freeze the assets of overseas governments or residents who had taken action, or had intent to act, against the UK's economy, or any special act which constituted a threat to the life or property of a national or resident of the UK.[58]

---

[57] This act was passed in 1974 one year prior to the Prevention of Terrorism Act in 1974 by Westminster Parliament in response to two bomb attacks in Birmingham, England. The Emergency Provision Act authorised police inside Northern Oreland to stop and investigate individuals without prior judicial notice or order.

[58] M. Charvat, 'A study of UK anti-terror law, legal aspect of combating terrorism', *In Centre of Excellence Defence Against Terrorism.* IOS press, Ankara, Turkey, 2008, p. 109.

It must be emphasised that there is no need to be connected to a proscribed organisation. Furthermore, Section 57 deals with being caught red-handed with possession of explosives or electronic timers. Therefore, the law is intended to cover people involved in activities which do not need actions directly related to terrorism and cyberterrorism. This Section can be applied for cyberterrorism cases wherein suspects are arrested because of suspicious items that they possess. Subsection (3) of the Act covers proof of 'possession'.

That is, if it can be proved that an article: "(a) was on any premises at the same time as the accused, or (b) was on premises of which the accused was the occupier or which he habitually used otherwise than as a member of the public", then the court could assume that the accused possessed the document or record, unless the accused proves that he did not know of its presence on the premises or that he had no control over the premises.[59]

In Section 57(3), if the defendant raises a defence, the burden of disproving the defence is on the prosecution and if there is a reasonable doubt it remains on the prosecution to prove that as well as the proving of essential facts. Because of the equivocal situation of Section 57, Section 118 was added to the Terrorism Act to augment Sections 57(2) and 57(3). According to Section 118, if evidence is adduced which is sufficient to raise an issue, the court "shall treat it as proved unless the prosecution disproves it beyond reasonable doubt". This formula was intended to be merely declaratory.[60]

Hence, after the raising of a defence by a defendant, the burden of proof remains on the prosecution. In addition, the prosecution must prove reasonable suspicion of the item relevant to Section 57. Some commentators have said that Section 57 is likely to trivialise the burden on the prosecution, particularly due to the fact that there is a need of indirect proof of a terrorist purpose.[61] It must be added that

---

[59] Section 57(3) Terrorism Act 2000.

[60] Section 57 Terrorism Act 2000.

[61] C. Walker, 'Cyber terrorism: Legal principle and law in the United Kingdom', *Penn State Law Review* (2006) 110, p. 651.

at the end, the burden of proof is shifted onto the prosecution. After 2000, the usage of Section 57 has markedly increased.[62]

The controversial nature of Section 57 is illustrated in the case of Brahim Benmerzouga and Baghdad. Two Algerian-born men filed an official request for political asylum in Britain. Together with Kamel Daoudi, a Franco–Algerian terrorist suspect, they were arrested on 25 September 2001 for alleged ties to al-Qaeda and for helping to finance the organisation following the September 11 attacks. Brahim Benmerzouga admitted to one charge of conspiracy to defraud by manufacturing or using false bank cards and account details, as well as three charges of possessing false passports. According to the prosecutor Mr Mark Ellison, the two Algerians were linked to a network of persons covering Europe who shared common interests in perpetuating Islamic extremism.

He added that the defendants were in possession of a large collection of "shocking and heavily religious" videos that included 19 copies of a film that promoted Osama bin Laden's speeches. Both the Algerians received 11-year prison sentences.[63] However, it is not clear why Section 58 was not invoked against Babar Ahmad, a computer analyst who worked at Imperial College, London. He had been accused of possessing material supporting terrorism, support of the Taliban and Chechen rebels, conspiracy to kill, money laundering, and conspiracy. He was under suspicion for raising money for terrorists through several websites. He was arrested by the British authorities in December 2003 but released after the commencement of extradition proceedings by US authorities.[64]

According to Section 59 of this Act, a person commits an offence if "(a) he incites another person to commit an act of terrorism wholly or partly outside the United Kingdom, and (b) the act would, if

[62] L. Carlile, Home office, Report on the Operation in 2002 and 2003 of the Terrorism Act 2000. Available at: http://tna.europarchive.org/20100419081706/http:/security.homeoffice.gov.uk/news-publications (20 Aug 2013).

[63] Global Jihad, Brahim Benmerzouga and Baghdad Meziane, 2010. Available at: http://www.globaljihad.net/view_page.asp?id=1944 (1 Sept 2013).

[64] C. Walker, 'Cyber terrorism: Legal principle and law in the United Kingdom' (2006) 110, Penn State Law Review, p. 651. Available at: http://news.Findlaw.com/cnn/docs/ahmad/usahmad/72804cmp.pdf (10 Jan 2011).

committed in England and Wales, constitute one of the offences listed in subsection (2)". This Section potentially requires the UK to protect every government in the world. According to subsection (4), it is expressly "immaterial whether or not the person incited is in the United Kingdom at the same time of the incitement".[65] Sections 59–61 turn certain offenses into universal crimes when they are recognised as such elsewhere.[66] These terrorist activities committed outside the UK are treated the same way as offences committed inside the country.

Another important terrorist use of cyber space is for propaganda. Other offences involving the internet are more acceptable to the notion of violence rather than politics. These include Section 54; under this section a person commits an offence if he: "provides instructions or training in the making or used of (a) firearms, (b) radioactive, material or weapons design or adopted for the discharge of any radioactive material, (c) explosives, or (d) chemical, biological or nuclear weapons".[67]

In a similar way, a person who uses the internet for terrorist propaganda is also attempting to launch terrorist activities. Existing laws related to terrorism offenses must be used for cyber terrorist activities. Terrorism offenses were considered in part 6 of the Terrorism Act of 2000. As previously mentioned, training and providing instructions for terrorist activities are considered as offenses, and a person who commits such crimes shall be liable for conviction to incitement or imprisonment. Spreading terrorist propaganda via the internet can be interpreted as providing instructions and materials for terrorist purposes because people are taught various methods of launching cyber attacks.

It is an offence under Section 54(2) to receive instruction or training or, under Section 54(3), to invite another to receive instruction or training contrary to subsections (1) or (2), even if the activity is to take place outside the UK.[68] Interpreting Section 54(4),

[65] Section (59) Terrorism Act 2000.
[66] Section (62) Terrorism Act 2000.
[67] Section (54) Terrorism Act 2000.
[68] Section 54(2) Terrorism Act 2000.

"instruction" and "invitation" can be performed through general activities such as the use of pamphlets or the internet, or "addressed" to one or more specific persons. "In this way no identifiable recipient is needed for the offense to be committed".[69] Cyber terrorist activities or ancillary cyber activities are thus covered by these provisions. Sulayman Balal Zainulab Idin, a chef from Greenwich, south-east London, was charged under Section 54 in October 2001, arising from his activities in running an enterprise called Sakina Security Services which had advertised on the web training for Muslim recruits to prepare for "the ultimate jihad challenge", some of which was to occur at a facility called "Ground Zero" in Marion, Alabama, in the US.[70]

C. Malaysia

Under Section 7 of Malaysia's Computer Crime Act 1997, anyone who abets the commission of an offense under the Act, or does any act preparatory to or in furtherance of an offense, is guilty of the substantive offense. The act can impose the same penalty for abetting or attempting to commit an activity as it does for a substantive violation. Actions aimed at preparing for or promoting a terrorist activity are punishable under the act with a half the maximum prison term, the full fine, or both for a substantive offence.[71] This provision treats both the attempt and the preparation as offences. Although Section 511 of the Penal Code does not consider preparation to commit an offence a crime, it is considered as such under Section 7(2) of the CCA 1997. Thus it is clear that based on the current criminal law there is difficulty in differentiating between the preparatory stage and the actual attempt. [72]

---

[69] C. Walker, 'Cyber terrorism: legal principle and law in the United Kingdom', *Penn State Law Review* (2006) 110, p. 658.

[70] C. Walker, 'Cyber terrorism: Legal principle and law in the United Kingdom', p. 658.

[71] D. L. Beatty, 'Malaysia Computer Crime Act 1997 gets tough on cyber-crime but fails to advance the development of cyber law', *Pacific Rim Law and Policy Association* (1998) 7(2), pp. 363–364.

[72] D. L.Beatty, 'Malaysia's Computer Crimes Act 1997 gets tough on cyber-crime but fails to advance the development of cyber laws', pp. 363–364.

Sections 130D to 130L treat ancillary terrorist activities as criminal acts and include committing terrorist acts, providing explosives, recruitment for terrorist groups, providing training, instructions or facilities to terrorists, supporting and directing terrorist activities, knowingly inciting, promoting or soliciting property for the commission of terrorist acts, soliciting and giving support to terrorist groups for the commission of terrorist acts, providing services for terrorist purposes, dealing with terrorist property and criminal conspiracy.

Finally, as Denning, Flemming, Stohl, and Weimann have said: "Terrorist use of computers as a facilitator of their activities, whether for propaganda, recruitment, data mining, communication, or other purposes, is simply not cyberterrorism". Considering the types of activities that further cyber activities, it is concluded that "it is important to investigate the conditions under which terrorists would choose to employ digital means to advance their cause over conventional methods".[73] Denning suggests that:

> To understand the potential threat of cyberterrorism, two factors must be considered: first, whether there are targets that are vulnerable to attack that could lead to violence or severe harm, and second, whether there are actors with the capability and motivation to carry them out.

To determine motivation we have to ask not simply "if they desire to cause harm and exploit fear but also if the investments needed to create the event are more or less 'costly' than traditional means of terror".[74] Considering such elements provides useful insights into the process of understanding why terrorist organisations choose cyber space to further their activities, which groups are most likely to do so,

---

[73] M. Stohl, 'Cyber terrorism: A clear and present danger, the sum of all fears, breaking point or patriot games?', *Crime Law and Social Change* (2007) 46, p. 232. Available at: www.springerlink.com/index/Y81 6117WW6058JP7.pdf (3 Jul 2012).

[74] D. E. Denning, Cyber Terrorism, Testimony before the Special Oversight Panel on Terrorism Committee on Armed Services, U.S. House of Representatives, 2000. Available at: http://www.cs.georget own.edu/~denning/infosec/cyberterror.html (3 Jul 2012).

and the conditions under which it would make sense to use digital rather than conventional tools to accomplish their ends.

In fact, using the internet to deploy terrorist activities gives benefits to terrorists through desired outcomes, with the probability that such action will bring about the desired state of affairs and save on the costs of engaging in the action. These costs include the response costs which might be incurred by target groups, and the production costs which encompass the costs of taking the action regardless of the reactions of others. Thus, a cyber disruption attack can be carried out with less effort, have the same chances of detection, give the same results and have a greater chance of success than an attack by conventional means such as armed attack or bombing, or a direct conventional attack on infrastructure that bypasses the infrastructure control systems. Given the above, it is believed that cyberterrorism would be a rational choice for terrorist groups, rather than traditional methods.[75]

In a similar context, the ease, low cost, speed, and anonymity of the internet as well as the lack of an international convention on cyber crime makes the cyber terrorist attack the most suitable choice and because of this, the degree of destruction increases.[76]

3.5 LEGAL RESPONSE ACCORDING TO COMPUTER CRIME STATUTES

In responding to internet-based attacks, three approaches exist: first, applying cyber crime provisions to terrorist internet-based acts; second, applying terrorist attack legislation to terrorist cyber attacks; and third, enacting new legislation regarding terrorists' use of the internet. Of these approaches, the implementation of cyber crime provisions to terrorist cyber attacks is the best approach in cases where

[75] M. Stohl, 'Cyber terrorism: A clear and present danger, the sum of all fears, breaking point or patriot games?' *Crime Law and Social Change* (2007) 46, p. 232. Available at: www.springerlink.com/index/Y81 6117WW6058JP7.pdf (3 Jul 2012).
[76] Son Than Dand, The prevention of cyber terrorism and cyber war, Issue Brief for the GA First Committee: Disarmament and International Security (DISEC), ODUMNC, 2011.

specific legislation dealing with terrorist use of the internet does not exist. With the increasing attacks that are planned to decrypt, distort, and disrupt communication systems, the problems facing the investigation of these attacks have also increased.

However, these types of offences occur through the internet and computers and the process of their investigation and prosecution is completely different from other types of crimes.[77] Consequently, in using cyber crime legislation and non-internet terrorism legislation to address terrorist use of the internet, several problems arise.

The best legal response to internet-based attacks in the current situation is criminalising the relevant acts, which is done by the Convention on Cybercrime. On this basis, the Convention on Cybercrime lists nine offences and member countries have agreed to adopt the necessary measures in their legislation. In doing so, the main objective of the Convention on Cybercrime as set out in its preamble is to pursue a common policy to protect society against cyber crime via international cooperation and national legislation.[78] Although it does not specifically mention cyberterrorism but rather the applicability of cyber crime provisions to terrorist-related acts, it contains provisions that can be used for addressing offences such as the misuse of the internet, and investigation, and prosecution. Computer technology plays a vital role in the commission of acts of terrorism, thereby countering this with computer-related legislation would be the best method against cyberterrorism. Mostly terrorists inflict systematic damage on target systems such as power supplies, electrical grids, critical systems, and infrastructure.[79]

It is necessary to overview and critique legislation used to prosecute cyber attacks as terrorist acts. Most computer offences are broad enough to include physical attacks to electronic systems and

---

[77] *Countering the Use Of The Internet For Terrorist Purposes-Legal And Technical Aspects* United Nation Counter Terrorism Implementation Task Force (CTITF), CTITF Publication Series, United Nation, 2011, p. 6.

[78] K. Geers, 'The challenge of cyber-attack deterrence', *Computer Law and Security Review* (2010) 26, p. 300.

[79] S. W. Brenner, A light speed attribution and response to cyber-crime/terrorism/warfare', *The Journal of Criminal Law and Criminology* (2006) 97, p. 400.

other infrastructures.[80] Offences must be successfully prosecuted in order to be solved. A proper investigation will lead to a better prosecution.[81] Much domestic legislation related to cyberterrorism has been passed since the September 11 attacks to counter transnational threats. One issue that was often overlooked by lawmakers was how such legislation would apply to new threats such as cyberterrorism. Therefore, they have now enacted a definition of terrorism that includes cyberterrorism, as well as specific provisions relating to attacks against electronic systems and different infrastructures. This research provides an overview and critique of legislation that is used to prosecute cyber attacks as acts of terrorism in three different countries. The methods used by cyber terrorists to launch their attacks vary. The first step in launching an attack is gaining unauthorised access to systems.

3.5.1 Unauthorised Access

'Unauthorised access' or 'access without authorisation' is to act or access without a lawful entitlement to be involved in the relevant conduct. The primary issue in this section is the broad concept of 'access'. Different jurisdictions give different meanings to the notion of 'access'. The definition of 'to access' seems problematic since 'it may be thought to encompass obtaining physical access to a computer, or extend to obtaining a hard copy of data stored in a computer'. Unauthorised access is a broad term and each jurisdiction has varied definitions and terminologies for it, but all of them have the same essential concepts. All jurisdictions' definitions are aimed at both outsiders gaining access and insiders who exceed authorised access. [82]

---

[80] K. Hardy, 'WWWMDs: Cyber-attacks against infrastructure in domestic Anti-Terror Law', *Computer Law and Security Review* (2011) 27, p. 153.

[81] S. Perumal, 'Digital forensic model based on Malaysian investigation process', *International Journal of Computer Science and Network Security* (2009) 9(8), p. 43.

[82] J. Clough, *Principles of Cyber-crime*, 1st Edn, Cambridge University Press, United Kingdom, 2010, p. 62.

A. Unauthorised access in the US

The CFAA makes it unlawful for an unauthorised person to access a protected computer in order to obtain information, acquire something of value through fraudulent means, or damage the computer of another. It also prohibits the dissemination of malware, which can intentionally damage a protected computer. Furthermore, persons who suffer damage due to violations of the statute are allowed to sue for compensatory damages. Such damages include "any impairment to the integrity or availability of data, a program, a system, or information", and those that cause losses aggregating at least US$5,000 in value during any 1-year period to one or more individuals. The punishment for any violation of the CFAA is a fine and/or imprisonment, where the prison term is limited to: "(1) not more than one year for the first conviction; (2) five years if the offence was committed for the purposes of commercial advantage or private financial gain, in furtherance of any criminal or tortuous act, or if the value of the information obtained exceeds $5,000; and (3) ten years for the second conviction".[83] This Act has been amended several times and the last amendment was by the USA Patriot Act (2001) to address cyberterrorism attacks and enhance the penalties for such attacks.

Section 1030(a)(3) condemns unauthorised access into Federal Government computers whether they are used by the government or the government shares access with others. Such trespasses are prohibited only when they affect use by the government or use for governmental purposes. The committee reports provide a useful explanation of the distinctive, "affects-the-use" element of the trespassing ban:

> Trespassing in a computer used only part-time by the Federal Government need not be shown to have affected the operation of the government as a whole. The Department of Justice has expressed concerns that the present subsection's language could be construed to require a showing that the offender's conduct would be an exceedingly difficult task for Federal prosecutors. Accordingly, Section 2(b) will

[83] T. Mythri Raghavan, 'In fear of cyber terrorism: An analysis of the congressional response', p. 301.

make clear that the offender's conduct need only affect the use of the Government's operation of the computer in question. [84]

The language of the statute indicates that it covers government computers that are available to non-governmental users and non-public computers may include government computers that the government allows to be used for non-governmental purposes. Therefore, it does not cover "protected computers", but it covers non-public, federal government computers. Although, federal criminal law is presumed to apply within the US, in some circumstances it outlaws the offence whether committed within the US or elsewhere. Furthermore, when the criminal statute is silent, the court concludes that Congress intended to apply overseas misconduct because of the nature of the offence and the circumstances under which it was committed.[85]

In *United States v. Bowman*[86] the Supreme Court concluded that Congress must have intended the federal statute that prohibited fraud against the federal government to apply to fraud against the US committed abroad, particularly when the offenders were Americans.[87] It seems that while under paragraphs 1030(a)(2), (4), (5), and (7) the court concluded that Congress intended to grant extraterritorial jurisdiction, it meant to exclude extraterritorial applications in trespassing cases under paragraph 1030(a)(3).

For example, "Solar Sunrise" was a major cyber attack against the US wherein attackers gained unauthorised access to the power and computer control networks of the US Armed Forces, after which a threat to US national security was simulated. Solar Sunrise hit US Air Force, Navy, and Marine Corps' unclassified computer networks worldwide. The attackers gained system administrator privileges.[88]

---

[84] C. Doyle, Cyber-crime: An Overview of the Federal Computer Fraud and Abuse Statute and Related Federal Criminal Law, 97–1025, CRS Report for Congress, 2010, p. 5.

[85] C. Doyle, Cyber-crime: An Overview of the Federal Computer Fraud and Abuse Statute and Related Federal Criminal Law, p. 6.

[86] 260 United States Report 94, [1922], p. 98.

[87] *United States v. Bowman*, [1922] 260 U.S. 94, p. 98.

[88] A. M. Half, 'Cyber Power as a Coercive Instrument', Ph.D Thesis of Advance Air and Space, Maxwell Air Force Base, Alabama 2009, p. 19.

This attack originated from a variety of countries such as the United Arab Emirates, France, Taiwan, and Germany and was launched by two teenage boys from Israel and America, but it remains unclear if this attack was an individual attack or was supported by a terrorist group. The Solar Sunrise attack made use of system vulnerabilities and followed the same profile as all other attacks, i.e. "(1) probe the system; (2) exploit the system; (3) implant a sniffer program into the system; and (4) return later to gather data the sniffer program collected from the system. The sniffer programs targeted key ports and gathered hundreds of network passwords".[89]

Section 1030(b) also censures attempted intrusions and conspiracies to intrude. It is apparent that the unauthorised access must be intentional and inadvertent trespass is not intended. However, according to the view of the house report, anyone "whose initial access was inadvertent but who then deliberatively maintains access after a non-intentional initial contact" is assumed to have intention.[90] The unauthorised access of a government employee was accepted under this Act, but divided into two situations: in the first situation, the employee exceeds his authorised access and peruses data belonging to the department, which he is not allowed to look at. The Act delineates administrative sanctions instead of criminal sanctions for government employees. Thus, it avoids repetitions of prosecutions for every time an employee exceeds his authorised access.[91] In the second situation, the government employee trespasses into computers belonging to another department, because the individual who is authorised to access the data of certain computers in one department is not authorised to access them all. Section 1030(a)(3) does not cover this part and is limited merely to cases where the offender is completely outside the government. This section does not apply to cyberterrorism cases because it only covers government employees.

[89] Solar Sunrise, Global Security Organization. Available at: http://www.globalsecurity.org/military/ops/solar-sunrise.htm (11 Apr 2009).

[90] C. Doyle, Cyber-crime: An Overview of the Federal Computer Fraud and Abuse Statute and Related Federal Criminal Laws, 97–1025, Congressional Research Service for Congress, p. 3.

[91] 18. United States Code 1030(E) (7).

Normally, cyberterrorism is perpetrated by outsiders. Nevertheless, cyberterrorism can also be launched from the inside.

Furthermore, this section does not cover a foreign offender who has unauthorised access to a "protected computer" in the US. Extraterritorial jurisdiction covers only some crimes such as computer-related information acquisition, fraud, damage, and extortion offences under paragraphs 1030(a)(2), (4), (5), and (7). It excludes extraterritorial jurisdiction in simple trespassing cases under paragraph 1030(a)(3).

An attempt to violate any of the provisions under subsection (a) is a federal crime. However, the defendant must take substantial steps towards the commission of the offense, which confirms his criminal intent. The defendant's act must show criminal intent. Mere preparation does not constitute a substantial step toward final commission. Unauthorised access is the first step of launching a cyber terrorist attack. A perpetrator must gain unauthorised access to a secured government computer to obtain necessary information.

A method by which cyber terrorists launch their attacks is through unauthorised access, enabling them to obtain information by using the computer either as a target of attack, or as a tool. Section 1030(a)2) US Code concerns the prohibition against acquiring certain protected information by intentional unauthorised access, or excess of authorisation. That is:

> "(a) Whoever ... (2) intentionally accesses a computer without authorization or exceeds authorized access, and thereby obtains (A) information contained in a financial record of a financial institution, or of a card issuer as defined in Section 1602(n) of Title 15, 49 or contained in a file of a consumer reporting agency on a consumer, as such terms are defined in the Fair Credit Reporting Act (15 U.S.C. 1681 et seq.); (B) Information from any department or agency of the United States; or (C) information from any protected computer ... shall be punished as provided in subsection (c) of this section".[92]

[92] 18. United States Code 1030(a)(2).

This is more than a simple trespass in contrast with Section 1030(a)(3), and it covers a wider range of computers, and includes information of the federal government, financial information, and information acquired from a protected computer. The term "exceeds authorised access" must be interpreted narrowly to avoid turning the CFAA into a statute that inadvertently criminalises a wide range of innocuous activity. Due to many debates over the term "unauthorised access" and "exceeds authorised access" for "obtaining information" the Department of Justice made clear that "obtaining information" means mere observation of data and there is no need to copy or transport the information. This explanation was ultimately enacted as part of the Economic Espionage Act 1996 and was extended to cover "information obtained from federal computers and information secured by interstate or overseas cyberspace trespassing".[93]

The information used in Section 1030(a)(2) includes information stored in intangible form and it is enough merely to read it and it is not required that the information be copied or transported. Thus, although the theft of electronic information cannot be charged under traditional criminal statutes, it is prohibited in the same way as theft of physical items under Section 1030(a)(2)(c). The theft of trade secrets when accessed from a protected computer can be prosecuted under Section 1030(a)(2) in tandem with the Economic Espionage Act.[94]

The Identity Theft Enforcement and Restitution Act of 2008 expanded the implications of Section 1030(a)(2) by eliminating the requirement of forbidden access and redefining "protected computers" to include computers affecting interstate or foreign commerce while broadening the protected elements by creating a crime of "conspiracy" to commit a cyber crime. It defined a protected computer as "a computer located outside the United States that is used in a

---

[93] The Economic Espionage Act of 1996 (EEA), 18 U.S.C. §§ 1831–39, protects proprietary economic information by making certain types of trade secret misappropriation federal crimes.

[94] C. Decker, 'Cyber-crime 2.0: An argument to update the United States criminal code to reflect the changing nature of cyber-crime', *USC University Southern California* (2008) 81(5), p. 987.

manner that affects interstate or foreign commerce or communication of the United States". This amendment appears as paragraph 814(d)(1) of the USA Patriot Act (2001). The USA Patriot Act (2001) confirmed the concepts of unauthorised access and information acquisition from abroad from a protected computer.

The drafter of this legislation believed that Section 1030(a)(4) outlawed fraud by computer intrusion, which may be achieved by a cyber attack. The elements of this crime include: "knowingly and with intent to defraud; accessing a protected computer without authorisation, or exceeding authorisation; thereby furthering a fraud and obtaining anything of value (other than a minimal amount of computer time, i.e. more than $5,000 over the course of a year)".

The first element of this crime shows that the offender is conscious of the natural consequences of his action. The phrase "thereby furthering a fraud" shows that the usage of the computer is significant in this category and that prosecution is limited to cases where use of a computer is central in the criminal investigation. Furthermore, the language of this section shows that more than mere unauthorised use is required. The additional end — obtaining information — is obtained through unauthorised access. In *United States v. Czubinski*,[95] the appeal was overturned because the government had failed to prove that it was defrauded. The evidence did not show anything more than curiosity by Czubinski to view information about his friends and acquaintances and political rivals. They did not find any evidence that he printed out, recorded or used the information he browsed.[96]

Access, use, disclosure, interception of electronic communications is also prohibited according to the ECPA 1986 which is a law that protects the privacy right of people when they use telephones, computers, cell phones, or other means of electronic transmission of communication. According to the ECPA any intentional and attempted interception of electronic communication and the use of illegally obtained electronic communication are prohibited under the ECPA. Title II of the Act also prevents hackers from obtaining, altering, or

---

[95] 2 USTC 50, [1997], p. 622.

[96] *United States v. Czubinski*, [1997] 106 F.3d 1069 (1st Cir).

destroying certain stored electronic communications. It can also be used on anyone who "intentionally accesses without authorization a facility through which an electronic communication service is provided; or ... intentionally exceeds an authorization to access that facility; and thereby obtains, alters, or prevents authorized access to a wire or electronic communication while it is in electronic storage in such system shall be punished".[97] For the prosecution of computer-related electronic communication violations, the ECPA and Federal Wiretap Act 1968 work together and the effect of this conjunction allows prosecutors to adapt with changing technology. Thus, it can be used for suing cyber terrorists' cases.

The unauthorised access from a protected computer for sending spam email is prohibited under Controlling the Assault of Non-Solicited Pornography and Marketing Act of 2003 (CAN-SPAM). It covers spam that dominates another type of cyber crime, which is also a method of launching cyber terrorist attacks. Spam is defined in this Act as "any electronic mail message the primary purpose of which is the commercial advertisement or promotion of a commercial product or service (including content on an internet website operated for a commercial purpose)".[98] It prohibits: "the intentional sending of spam from a protected computer without authorisation;[99] the use of a protected computer to send spam with the intent to deceive recipients of its origin;[100] the use of a protected computer to send spam with materially false headings;[101] and the false representation of origin in a spam message".[102]

It criminalises the use of spam as a mass marketing tool and broadens the scope of prosecutable acts. If the spam crime includes the intent to cause damage to a protected computer, the CAN-SPAM Act joins with the CFAA. Spam prosecution carries up to a 5-year jail sentence, a fine, or both. Thus, spam, as a significant method used by

---

[97] 18 U.S.C. Section 2701(a)(2000).
[98] 18 U.S.C. Section 7702(2)(A).
[99] 15 U.S.C. Section 7704(b)(3).
[100] 15 U.S.C. Section 7704(a)(1)(C).
[101] 15 U.S.C. Section 7704(a)(2).
[102] 15 U.S.C. Section 7704(a)(1).

cyber terrorists bombarding a target and impairing and infecting infrastructure sites, can be prosecuted under this Act in the US.

Furthermore, the CAN-SPAM Act 2003 can be applied to phishing crimes. Generally, there is no specific statute for phishing; therefore, it is charged under the CFAA, the federal wire fraud statutes, the CAN-SPAM Act, and federal trademark law. For instance, a spoofed email that activates a phishing scam is prosecuted under the CAN-SPAM Act. The fraud that gives a phisher access to a protected computer by stealing the password is prosecuted under the CFAA. Any unlawful infringement on trademark, such as falsified email or copied websites, may be prosecuted under trademark law.

i. *Unauthorised access for launching federal crime of terrorism*

Generally, the federal crime of terrorism in paragraph 1030(a) US Code includes the paragraphs 1030(a)(1) and (a)(5) US Code. Section 1030(a)(5) US Code is the primary tool to investigate and prosecute hacking crimes. Furthermore, most of these crimes are against a "protected computer" and much debate exists over the precise definition of this term. Finally, Congress added the phrase "including a computer located outside the United States that is used in a manner that affects interstate or foreign commerce".[103]

Section 1030(a)(5) establishes three distinguishable means of damage on the basis of the offender's intent to damage: (A) intentionally causing damage without authorisation to a protected computer through a knowing transmission. Such transmission must be implemented with the installation of a destructive program, by inserting code, or other information. This covers most hacking crimes and even covers distributed denial of service (DDoS) attacks, and a hacker faces dual liability with regard to both the targeted system and the "zombie" system, since the attacker causes the transmission of information, packets, and code with the intent of harming both systems. Under this section of the Criminal Code, a hacker who infiltrates a system and in some way damages the system or the data on it faces liability.

---

[103] 18 U.S.C. 1030(e)(2).

Section 1030(a)(5)(B) of US Code is about recklessly causing damage to a protected computer by intentional unauthorised access. This section is applicable for hacking crimes in which damage is caused inadvertently. Section 1030 (a)(5)(C) of US Code is about causing damage and loss to a protected computer by intentional unauthorised access. This subsection was created for prosecutors to make a case if they were still unable to make a case under subsections (A) and (B). This is because it prohibits unauthorised access that causes damage negligently. If this provision did not exist, Congress would have implicitly accepted hacking into a computer or system so long as no damage occurred. Subsection (C) indicates that it covers both direct and indirect damage.

This feature shows that anyone who secures unauthorised access intentionally is punishable for the resulting damage regardless of whether they intended to cause it or it happened recklessly. This paragraph refers to recklessly and intentionally causing damage, ranging from serious damage to less serious damage treated as a misdemeanour. Although simple damage is treated as a misdemeanour, it requires proof of damage and of a loss.[104] Damage is defined as "any impairment to the integrity or availability of data, a program, a system or information".[105] This feature of computer damage offences was added in 1996 and amended by the USA Patriot and Homeland Security Acts. It emphasises the previous understanding that anyone who intentionally secures unauthorised access is punishable for any damage which results from that unauthorised access regardless of whether he intended to cause damage, or if it were recklessly done, it is no different as to whether he did so.

The definition of "damage" by the US Code is provided broadly, which results in the treatment of this definition in an ambiguous manner. The statute identifies damages that interfere with the integrity of a computer and provide specific foreseeable damages, such as "the cost of responding to an offense, conducting a damage assessment, and restoring the data, program, system, or information to its condition

[104] United States Code 1030 (a)(5)(C).

[105] United States Code 1030 (e)(8).

prior to the offense, and any revenue lost, cost incurred, or other consequential damages incurred because of interruption of service".[106] The courts have found loss which is "the natural and foreseeable result" of a violator's conduct. They have not required a loss of physical damages in the traditional way, so long as the US$5,000 threshold is met as a result of a violation of the US Code.

DDoS attacks are prosecuted under the CFAA as a part of the US Code since these sorts of attacks include unauthorised or excessive access to other computers which are well-suited to be prosecuted under the CFAA. The US$5,000 threshold requirement for damage must be met, but this damage is easier to allege in DDoS attacks than in a single hacking crime. In 2005, a man pleaded guilty to infecting thousands of computers through a "zombie computer" using a worm program that took advantage of computer vulnerability, then launching them in a DDoS attack. The DDoS attack impaired infected computers and he pleaded guilty and was convicted and sentenced to 10 years in prison according to Sections 1030(a)(5)(A) and (a)(5)(B).[107]

Unauthorised access is condemned in Article 1030 of US Code. Section 1030(a)(3) criminalises unauthorised access into federal government computers whether they are used by the government or the government shares access with others. The penalties for conspiracy to violate, or for violations or attempted violations of, paragraph 1030(a)(3) are imprisonment for not more than 1 year and/or a fine of not more than US$100,000 (US$200,000 for organisations) for the first offence and imprisonment for not more than 10 years and/or a fine of not more than US$250,000 (US$500,000 for organisations) for all subsequent convictions.[108]

The main intent of this clause is to make DoS attacks illegal as one of the tools of launching cyber terrorist attack. The requirement in this provision is that the guilty person must know his act is unauthorised

---

[106] 18 U.S.C. 1030 (e)(11).

[107] C. Doyle, Cyber-crime: An Overview of the Federal Computer Fraud and Abuse Statute and Related Federal Criminal Laws, 97–1025, *Congressional Research Service for Congress*, p. 64.

[108] 18 U.S.C. 1030(c), 3571.

when he does the act and such unauthorised acts must be in relation to a computer. In order to convict under this act, the effect of unauthorised act must be:

> a. to impair the operation of any computer; b. to prevent or hinder access to any program or data held in any computer: c. to impair the operation of any such program or the reliability of any such data: or d. to enable any of the things mentioned in paragraphs (a) to (c) above to be done.

Then the guilty person convicted of the offence may be sentenced to a maximum of 12 months imprisonment on summary conviction or 10 years on indictment.[109] The fault element is very important in this section. It is not intended to cover the offences occurring by unauthorised use but without any intention. Mere use without intention will not fall within this section.[110] This Act emphasises the point that impairment will be caused but it does not matter that the conduct is either intentional or reckless. However, for the prosecution, proving the intent of the offence is difficult. The prosecutor must also prove that the elements of the crime had been satisfied.

Section 1030(a)(5) of the US Code is the primary tool to simultaneously investigate and prosecute hacking crimes. It can cover hacking crimes and DDoS attacks which are the main part of cyberterrorism attacks. Hackers face liability with regard to both the targeted system and the 'Zombie' system since the attacker causes the transmission of information, packets, and code with the intent of harming both systems. In order to establish the transmission element of the intentional damage "the government must offer sufficient proof that the person charged is the same person who sent the transmission. Circumstantial evidence is sufficient to prove that the transmission occurred".[111] Such transmission must be implemented with the installation of a destructive program, by inserting code, or

---

[109] Section 36 The Police and Justice Act.

[110] J. Clough, *Principle of Cyber Crime*, p. 104.

[111] *United States v. Shea*, [2007] 9th Cir 493 F.3d , pp. 1110 and 1115.

other information. According to this subsection only the offences that knowingly or recklessly cause or attempt to cause death will attract the maximum penalty of life imprisonment. A cyber attack that causes or attempts to cause serious bodily injury will only attract a maximum penalty of 20 years' imprisonment.[112] Subsection 1030(a)(5)(A) of the US Code is punishable by imprisonment for not more than 10 years (not more than 20 years for a second or subsequent offence) and/or a fine of not more than US$250,000 (not more than US$500,000 for an organisation).[113]

Section 1030(a)(5)(A) and (B) of the US Code is about intentionally and recklessly causing damage to a protected computer by intentional unauthorised access. Subsection (C) of this part includes intentionally accessing to a protected computer without authorisation, and as a result of such conduct, causing damage and loss is liable under this section. It seems that subsection (C) was created for prosecutors to make a case if they were unable to make a case under subsection (A) and (B). This is because it prohibits unauthorised access that causes damage negligently. However, damage and loss require proof of damage.

The activities under subsection 1030(a)(5)(A) and (B) require "loss to or more persons during any 1-year period (and, for purposes of an investigation, prosecution, or other proceeding brought by the United States only, loss resulting from a related course of conduct affecting 1 or more other protected computers) aggregating at least $5,000 in value".[114] Both Sections 1030(a)(4) and 1030(a)(5)(A)–(C) require damages greater than US$5,000. However, the US$5,000 loss requirement is difficult to establish and investigate for prosecutors. The courts have found loss which is "the natural and foreseeable result" of a violator's conduct. They have not required a loss of physical damages in the traditional way, so long as the US$5,000 threshold is met as a result of a violation of the US Code.

---

[112] 18 U.S.C. (a)(5)(A)(i).

[113] 18 U.S.C. 1030(c)(4)(B), (c)(4)(C), 3571.

[114] 18 U.S.C. 1030(a)(5)(B)(i).

The attempt of disclosure of information is tracked through paragraph 1030(a)(1). The intent element of this crime must be fulfilled by the offender: (1) purposefully transmits or retains information that, (2) he has reason to believe could be used to the injure the US or benefit another country, and (3) that he has obtained through access to a computer that he knows he had no authority to access. The elements of this offence must be fulfilled in bad faith. Such crimes, according to these guidelines, are punishable by imprisonment for not more than 10 years (not more than 20 years for second and subsequent offences) and/or a fine of not more than US$250,000 (not more than US$500,000 for organisations).[115]

B. Unauthorised access in the UK

The UK uses the phrase "cause a computer to perform any function" that covers any manipulation. This phrase is considered broad as it cannot be subject to any limitation in the manner that a defendant performs an offence covered by the Act. In the UK, unauthorised access covers access to data and programs. Therefore, the focus is on the programs and data being accessed, even if that access is authorised. The access to the data and programs may be subsequent to an authorised access and not limited to initial access to the computer. It encompasses all the possible acts done on or to a computer. Put simply, any input to a computer with unauthorised access, if it is accompanied with relevant intent and causes that computer to function at some level, qualifies as unauthorised access.[116]

The *mens rea* of Section 1 of Computer Misuse Act 1990 has two elements; first, there must be "intent to secure access to any program or data held in any computer or enable any such access". The offender is not required to commit any damage or to have any intent to cause harm. This section criminalises the most basic level of hacking at once. The last part of the paragraph was amended in 2006 and criminalises activities beyond the use of hacking tools. For instance, if a person

[115] 18 U.S.C. 1030(c)(1), 3571.

[116] J. Clough, *Principles of Cyber-crime*, p. 62.

disables the control mechanism of a computer without impairing it or he prepares to use it in the future. Second, the offender must know that his act is unauthorised during his commission (*actus reus*).[117]

Unauthorised access to computer systems creates a huge opportunity for inflicting damage to computer data and programs. Section 1 of this Act is the "Offence of unauthorised access" which includes the offences of "hacking" and "cracking". Section 1(a) provides a broad definition of the commission of the offence. The basis of this Act is constructing the *actus reus* of an offender as "a computer to perform any function". Put another way, according to this provision, the perpetrator is not required to have his intention to be directed at any particular data, program, or data held in a particular program. The mere unauthorised access, without anything more, is an offence and even the hacker who inputs numbers at random to discover gateways to a computer system is rendered liable to prosecution. A hot debate has risen between the Scottish Law Commission and the Law Commission of England and Wales regarding the many instances that have happened through unauthorised access and which have caused computer owners to expend a huge amount of money to be certain that no damage has occurred. In this respect, mere unauthorised access becomes sufficient as the basis of liability and this approach is adopted in the Computer Misuse Act.[118]

Another kind of unauthorised access in the UK is through interception of data. As discussed previously, a cyber attack can occur during the transmission of data and with the unauthorised interception of communication. The Convention on Cybercrime also requires member parties to provide for offences which relate to "the interception without right, made by technical means, of non-public transmission of computer data to, from or within a computer system, including electromagnetic emissions form a computer system carrying such computer data".[119] The interception must be committed 'intentionally', without right. It applies to non-public transmission of computer data,

---

[117] C. Reed, *Computer Law*, 7th Edn, Oxford Publication, United Kingdom, 2011, p. 50.

[118] C. Reed, *Computer Law*, p. 50.

[119] Convention on Cybercrime, Article 3.

which focuses on the nature of the transmission that parties wish to be confidential.

The interception must be committed by 'technical means' and within a computer system, and the parties may be required to consider additional elements to be applied to computer systems remotely connected.[120] In an attempt to provide a single framework for the interception of data, the UK enacted the Regulation of Investigatory Powers Act 2000. According to this Act, intercepting any communication in the course of transmission by means of a public or private telecommunication system without lawful authority and intentionally in the UK is an offence that carries 2 years' imprisonment.[121] Such behaviour is unlawful according to this Act, unless in circumstances where an interception is under a warrant, or lawful operation of a service provider, or the consent of the operation controller is expressed.[122]

For the prosecution of unauthorised access in the UK, the prosecution must prove that the defendant has both the intention to secure access and knowledge that it was unauthorised. However, it is not required for the prosecution to prove that the defendant intends to access any particular data, or a program, or data held in any particular computer.[123] The basis of unauthorised access in the UK is the *actus reus* of an offender. According to this provision, the perpetrator is not required to have his intention to be directed at any particular data, program, or data held in a particular program or at a particular program or data. The mere unauthorised access, without anything more, is an offence and even the hacker who inputs numbers at random to discover gateways to a computer system is rendered liable to prosecution.

Therefore, in a situation where a cyberterrorism case happens in England which previously had the elements of a cyber terrorist act,[124] even when the terrorist cannot launch the attack, his mere unauthorised access would have sufficient potential to be prosecuted. A variety

[120] Convention on Cybercrime, explanatory reports, [55]. Accessed at http:// http://www.worldlii.org/int/other/COETSER/toc-C.html.

[121] Articles 1(1)–(2) Regulation of Investigatory Powers Act 2000.

[122] Articles 1(6) and 3(1)(2). Regulation of Investigatory Powers Act 2000.

[123] Section 1(2) Computer Misuse Act.

[124] It was described in Chapter 2.

of conditions for unauthorised access must be satisfied. For instance, access must be attempted or obtained, the access must be unauthorised and the person must know that is the case. However, proving these requirements and their application is somewhat complex. For example, it is very difficult to prove that they had actual knowledge that access was unauthorised. As to its legal effect, the installation of a security system or the message "unauthorised access to this system is illegal" on the computer system might be sufficient to support the idea that any further offences might be interpreted as knowledge. Some scholars believe that it is a broad formulation which causes even the turning on of a computer to be a necessary act.

According to the UK Computer Misuse Act, the mere intention of a perpetrator for unauthorised access makes the person liable for prosecution. The broad phrase "cause a computer to perform any function" encompasses all possible acts with a computer. For instance, any input into a computer by a person not having authorised access to it, if accompanied with relevant intent, that causes that computer to function at some level, qualifies as unauthorised access.

C. Unauthorised access in Malaysia

Section 3 of the Computer Crime Act 1997 of Malaysia stipulated that an offence of unauthorised access is committed if a person causes a computer to perform any function with intent to secure access to any program and data held in any computer and that access is unauthorised and the person knows that at the time he caused the computer to perform the function that was the case. Here, intent is not required to be directed towards any particular program or data of any kind or a particular computer. The Act just criminalises any intentional access to any computer without authorisation.

Unlike Malaysia, many countries, such as the US, have criminalised unauthorised access to computers; this applies only to computers that contain national security data. Under the CCA 1997, the offender is liable to a fine not exceeding RM50,000 or to imprisonment for a term not exceeding 5 years, or both. According to Section 3 of the CCA 1997, the offence of unauthorised access to computer material is a required element. A person shall be guilty of this offence if: (a) he causes a computer to perform any function with intent to secure access to any program or data held in any computers; (b) the access he intends to secure is unauthorised; and (c) he knows at the time when he causes the computer to perform the function that is the case. Hackers perform different functions to secure access to a program or data in any computer. The phrase "to perform a function" somehow seems ambiguous. According to the Economic Espionage Act, the notion of "function" includes logic, arithmetic, deletion, storage, retrieval, communication, and telecommunication to, from, or within a computer. From this definition, the offense requires not just the obtaining of access to any hardware involved, but also requires an attempt "to perform a function" at the very least.

The first requirement of Section 3 is satisfied when efforts for committing the offense are successful, even though the hacker may not be aware of the identity of the victim. The second element is for the commission of unauthorised access that is not confined to just someone who has no access rights, but person who exceeds his authorisation as well. A would-be hacker must be aware that his access is unauthorised for him to be punished under this provision.[125] The determined sentence for this offence goes beyond the OECD recommendation and that of the Council of Europe.[126] Although

---

[125] A. Abdul Rahim *et al.*, 'Theft of information: Possible solutions under Malaysia law' (2003) 3, *Malaysian Law Journal*, p. xc.

[126] The OECD recommended to member states that they adopt laws that would consider five activities as an offense in 1986. These activities include: inputting or altering data or programs with the intent to illegally transfer funds or other items of value; inputting or altering or altering data with intent to disrupt the functioning of a computer; the infringement of the exclusive right of the owner of a protected computer program with the intent of commercial exploitation; and the access of a computer system to infringe security measures or for other dishonest or harmful means. This long list served as a basis for harmonising computer crime laws among the member states of the OECD. The Select Committee of Experts on Computer-related Crime of the Council of Europe and the European committee on crime problems prepared recommendation No. R(89)9 on computer crime problem in September 1989. Several offenses were considered that include: computer fraud, computer forgery, damage to computer data or computer programs, computer sabotage, and unauthorised access by infringing security measures.

unauthorised access exists in their mandatory lists, such an act is only included as an offence if security measures[127] were infringed.[128]

Section 4 of the CCA 1997 is that if the offence is carried out with intent to commit fraud, dishonesty, or to cause injury as defined in the Penal Code, or to facilitate the commission of such an offence whether by himself or another person, he shall be liable to a fine not exceeding RM150,000 or to imprisonment for a term not exceeding 10 years, or both.[129]

Offences under this section can be categorised into unauthorised access and the act of committing or facilitating activities involving fraud, dishonesty, or causing injury. An accused cannot be convicted unless the element of unauthorised access is proven. For instance, if after accessing a website in an authorised manner a person spreads derogatory material or words, he is not liable under Section 4 of this act, although he has committed a further offence as the element of unauthorised access is lacking.[130]

For an offence under this act it is immaterial whether the offence is committed at the same time as when the unauthorised access occurs or on any future occasion.[131] Under the same section, charges will more likely result in conviction if the intention can be proved. In *R v. Thompson* as soon as the perpetrator obtained access to the data with the intention of modifying it, an ulterior intent is implied although the repetition of a conduct and they could not be prosecuted on the basis of theft or the obtaining of property.[132]

---

[127] The security measures deployed by the computer owner such password protections.

[128] D. L. Beatty, 'Malaysia Computer Crime Act 1997 gets tough on cyber-crime but fails to advance the development of cyber law', *Pacific Rim Law and Policy Association* (1998) 7(2), p. 87.

[129] Section 4 The Computer Crime Act.

[130] A. Abdul Rahim *et al.*, *Cyber-Crimes: Problem and Solutions Under Malaysian Law*, Jenayah Berkatikan dengan Komputer, Perspektif Undang-Undang Malaysia, Dewan Bahasa dan Pustaka, Kuala Lumpur, 2004.

[131] A. Abdul Rahim *et al.*, 'Theft of information: Possible solutions under Malaysian law', *Malaysian Law Journal* (2003) 3, p. ci.

[132] N. Abdul Manap, Cybercrimes: Problems and solutions under Malaysian Law, 2 '*lawyer*', Khorasan Bar Association (2012).

Cyber terrorists use computers as tools to target other computers. These people use modern and high-tech methods to launch their cyber attacks. Like other cyber crime, their first action is to gain unauthorised access to a computer system with the intention to commit a further offence. The target and intent of cyberterrorism distinguish it from other types of cyber attacks. The Computer Crime Act of 1997 can cover cyberterrorism attacks, specifically the initial stage of gaining unauthorised computer access.

The purpose of Section 6 is to address wrongful communication. A person shall be guilty of an offence if he communicates to a computer directly or indirectly a number, code, or password to any computer, which he is not authorised to communicate. The penalty for this offence is a maximum fine of RM25,000 or 7 years' imprisonment or both. Section 6 is not specific as to whether an unintentional communication of a password is criminalised or not. The Malaysian government wants to create strict liability, but they do not state anything clearly about the intent. In the theory of strict liability, the imprisoning of a defendant by a judge in trial may face some difficulty because the applicability or non-applicability of the *mens rea* is left to the discretion of the judiciary.[133] Strict criminal liability is being criticised by legal scholars because it does not serve the goal of punishment as a deterrent and rehabilitative measure.

By the same token, as soon as terrorists gain unauthorised access to computer systems, they can alter information, modify programs, obtain passwords, and monitor information being used or stored. The Computer Crime Act of 1997 covers all of these steps because they are mutual stages between methods for cyber attacks and cyber crime attacks. Cyber terrorists' methods are similar to cyber crime methods; using the cyber crime act to convict cyber terrorists is thus plausible. Cyber terrorists execute cyber crime methods swiftly and move into cyberterrorism attacks that threaten vast numbers of potential targets. These methods also introduce malicious codes that interrupt operations on a global scale.

---

[133] D. L. Beatty, 'Malaysia computer crime act 1997 gets tough on cyber-crime but fails to advance the development of cyber law', p. 355.

3.5.2 Exceeding Authorised Access

Exceeding authorised access happens where the defendant is authorised for a specific purpose, but he or she exceeds that authorisation. Determining the scope of the authorisation is the difficult part of the issue.

A. Exceeding authorised access in the US
"Exceeds authorised access" in the US is defined as "to access a computer with authorization and to use such access to obtain or alter information in the computer that the accesser is not entitled so to obtain or alter".[134] According to this definition, the distinction between "without authorisation" and "exceeding authorised access" emerges, despite the narrowness of the distinction, it is not invisible. It is an authorised access and such authorisation may be granted in two ways: by code (password) and by contract.[135]

The contractual limitation only binds the parties to the contract, and the limitation "depends upon the express or implied terms of limitation".[136] Another limitation on the scope of unauthorised access to information on computers is the fault element. According to this general rule, a defendant must know that his or her conduct was unauthorised.

B. Exceeding authorised access in the UK
If the conduct of the defendant is within the scope of authorisation, the act is authorised, but if the conduct of the defendant is beyond the scope of authorisation, the act is unauthorised. An analogy would be the law on burglary that requires the defendant to enter into the property as a trespasser. In the case of *Department of Public Prosecutions v. Murdoch*, the defendant was an employee of the State Bank of Victoria in the ATM section. He transferred to another section but he retained his earlier access to the computer controlling

---

[134] 18 U.S.C. 1030(e)(6).

[135] O. S. Kerr, 'Cyber-crime's scope: Interpreting 'access' and 'authorization' in computer misuse statutes', *New York University Law Review* (2003) 78, p. 1596.

[136] J. Clough, *Principles of Cyber-crime*, p. 77.

the ATMs. The functions of the system were organised in such a way that whenever a customer attempted to withdraw money from an ATM and it had insufficient funds, or the system was "on host" and able to communicate with the main computer, the transaction would be declined. However, if the system was "off host" and was unable to communicate with the main computer, the ATM would allow the customer to withdraw $200 without noticing if there were sufficient funds in the account. The defendant used his access to transfer the ATM to "off host"; thereby, he could withdraw money from accounts without sufficient funds in the accounts. It is clear that he was not authorised to do so. He was convicted as a "computer trespasser" under section 9A of the Summary Offences Act 1966 (Vic) which has now been repealed. According to that Section, it is an offence to "gain access to, or enter, a computer system or part of a computer system without lawful authority to do so". The question that arises here is that, although he had permission to access the ATM previously, the scope of the access is important. This was also seen in the case of *Barker v. R.*[137] In such a situation the defendant had exceeded the permission that was granted to him.

Section 2 of the Computer Misuse Act 1990 explains unauthorised access with a special intent to commit or facilitate commission of further offences. It defines further offences as ones for which the sentence is fixed by law. The access and the further offences do not need to be committed at the same time; also, it does not matter if the further offence was impossible. For instance, a hacker uses unauthorised access to a computer to obtain information to enable a later offence of cyberterrorism, but the further offence would not be committed until the information is used.

The interpretation of Section 17(5) was considered in detail in *Department of Public Prosecutions v. Bignell.* In this case, two police officers who gained access to the police national computer via an operator for personal purposes were convicted under Section 1 of the Computer Misuse Act with the accusation of unauthorised access for unauthorised purposes. The conviction was upheld by Bow Street

---

[137] *Barker v. R*, [1983] 153 CLR 338, p. 348.

Magistrates' Court and was confirmed by the Queen's Bench Divisional Court, even though the purpose of the access was not authorised. However, the court failed to make a distinction between the two entitlements of accessing computer material or controlling access to such material. These entitlements are subject to them being for job purposes only. Therefore, this decision appears to be wrong.

They appealed against their conviction to the Crown Court and the decision was dismissed by further appeal to the Division Court. This case concluded that "control access" in Section 17(5) did not refer to individuals authorised to access the system. This authority was given to the Commissioner for police purposes only. Lord Hobhouse pointed out that the computer operator did not exceed his authority, thus he did not commit an offence. What is more, he added that the concept of authorisation needed to be refined. The consequence of impairment by unauthorised access was damage. According to the Criminal Damage Act 1971, this kind of damage is proved when tangible property has been damaged and it is not necessary that the damage be tangible. However, the difficulty of the situation is in identifying the tangible property that has been damaged (under the Criminal Damage Act 1971 "property" means property of a tangible nature, whether real or personal).

This kind of crime happens through the manipulation of computer programs such as the insertion of a logic bomb to cause a computer to act in the desired way of the perpetrator rather than of its owner and at the same time it generates computer viruses and sends repeated threats to the computer owner. Until 1980, these cases were prosecuted under the Criminal Damage Act 1971. In the case of *R v. Whiteley* a computer hacker accessed a computer network and deleted files. This was detected and he was convicted for causing criminal damage. However, after being appealed on the grounds that the conduct of the perpetrator had not led to any tangible damage to the victim's computer, the Lord Chief Justice rejected this contention:

> What the act requires to be proved is that tangible property has been damaged, not necessarily that the damage itself should be tangible. There can be no doubt that the magnetic particle upon the metal discs were a part of the discs and if the appellant was proved to have

> intentionally and without lawful excuse altered the articles in such a way as to cause an impairment of the value or usefulness of the disc to the owner, there would be damage within the meaning of Section 1. The fact that the alteration could only be perceived by operating the computer did not make the alterations any the less real, or the damage, if the alteration amounted to damage, any less within the ambit of the Act.

Thus, a cyber terrorist may collect necessary information by exceeding his/her authorised access. Then they can alter, delete, or remove the infrastructure data of an organisation.

Unauthorised access with a special intent to commit or facilitate the commission of further offences is explained in Section 2 of the Computer Misuse Act 1990. The access and the further offences do not need to be committed at the same time; also, it does not matter if the further offence was impossible. A cyber terrorist who uses unauthorised access to a computer to obtain information to enable a later offence of cyberterrorism, the further offence would not be committed until the information is used. Prosecution under this provision is rare since most of the time there will be a prosecution for a further offence, rather than unauthorised access. This section covers more offences that are serious and includes hacking attempts with intention to do harm. Unauthorised access happens when: "(a) he is not himself entitled to control access of the kind in question to the program or data: and (b) he does not have consent to access given to him of the kind in question to the program or data from any person who is so entitled".[138] Under this provision, the accused must have entered into the system of the victim organisation and if the accused is an employee of the organisation, then the burden is upon the prosecution to show that the accused knew that such access was unauthorised. This happens when an authorised employee accesses the system for unauthorised use. According to case *DPP v. Bignell* and of *R v. Whiteley*, which is discussed in the previous chapter, the consequence of impairment by unauthorised access must be damage. It must be proved that tangible property has been damaged and it is not necessary for the damage to be tangible. However, identifying the tangible property that has been damaged can be difficult.

[138] Section 17(5) Computer Misuse Act 1990.

This kind of crime happens through the manipulation of computer programs such as the insertion of a logic bomb to cause a computer to act in the desired way of the perpetrator rather than of its owner and at the same time it generates computer viruses and sends repeated threats to the computer owner. Until 1980, these cases were prosecuted under the Criminal Damage Act 1971. In the case of *R v. Whiteley* a computer hacker accessed a computer network and deleted files. This was detected and he was convicted for causing criminal damage. However, after being appealed on the grounds that the conduct of the perpetrator had not led to any tangible damage to the victim's computer, the Lord Chief Justice rejected this contention.

C. Exceeding authorised access in Malaysia

The first requirement as per Section 3 is satisfied when efforts at committing the offense are successful even though the hacker may not be aware of the identity of the victim.[139] The second element is for the commission of unauthorised access that is not confined to anyone who has no access rights, but also to persons who exceed their authorisations. A would-be hacker must be aware that his access is unauthorised for him to be punished under this provision.[140] In such cases the employee has gone beyond his authority by exceeding the parameters or specifications of his work. Furthermore, the offender must be aware that access is unauthorised for the offence to be deemed to have been committed. Under this provision the acts must be intentional to be subject to penalties.[141]

3.5.3 Misuse of Devices

Article 6 of the Convention on Cybercrime obligates member states to adopt the necessary measures to establish criminal offences under

---

[139] The definition of unauthorised access is specifically given in Section 2 (5) Computer Crime Act 1997.

[140] A. Abdul Rahim *et al.*, 'Theft of information: Possible solutions under Malaysian law', *Malaysian Law Journal* (2003) 3, p. xc.

[141] A. Abdul Rahim *et al.*, 'Theft of information: Possible solutions under Malaysia law', p. xc.

their domestic laws. Under this article, they must criminalise the possession of "hacking tools" to combat against the growing market in dubious material.[142] Section 37 of the Police and Justice Act 2006[143] inserted a new Section 3A into the Computer Misuse Act 1990. This section covers three new kinds of offences: making, adopting, supplying, or offering to supply. Adopting offences are those regarding articles which may be used in the commission of an offence under Sections 1 or 3 of the Computer Misuse Act 1990. The first offence in Section 3A is when someone "makes, adapts, supplies or offers to supply any article intending it to be used to commit, or to assist in the commission of, an offence under Section 1 or 3".[144] Such terms are capable of encompassing a broad range of conduct. The terms "make" and "adapt" include the creation of tangible data, but they are not limited to intangible data and may include tangible data as well.

The second offence is "to supply or offer to supply any article believing that it is likely to be used to commit, or to assist in the commission of, an offence under Section 1 or 3".[145] The requirement in this provision is that there must be a reckless "supplying" or "offering to supply", since it is unusual that a person would recklessly "make" or "adapt" an article. For instance, a person produces and distributes some information without the intention of using it, but he acknowledges that it may be used by others. Therefore, according to the language of the section, the prosecution must prove an actual belief and a probability is insufficient. The third offence "is to obtain any article with a view to its being supplied for use to commit, or to assist in the commission of, an offence under Section 1 or 3".[146] This is

---

[142] Article 6 of the Convention on Cybercrime.

[143] It is an act extending police powers and granting the Home Office greater control over operational policing. The Computer Misuse Act 1990 under the section called 'Miscellaneous Part 5 Computer Misuse amendments' has amended by The Police and Justice Act 2006. Sections 35–38 Police and Justice Act 2006 will amend the Computer Misuse Act 1990 when in force. The Act was given Royal Assent and accepted into UK law on Wednesday 8 November 2006.

[144] Section 3 Computer Misuse Act 1990, (Act 3A).

[145] Section 3 Computer Misuse Act, (Art 3A (2)).

[146] Section 3 Computer Misuse Act, (Art 3A (3)).

aimed at persons who obtain such articles for the purpose of providing them to others to commit an offence, since according to this Act possession is not an offence even if accompanied by the intent to commit an offence.

Although this amendment to the Act provides comprehensive protection against DoS attacks, there are some vague areas. First, there may be a problem in the dual interpretation of the nature of an "article"; second, the words "believing that it is likely" are vague in nature as it is difficult for one person to accurately ascertain the intention of another in all situations; third, the section does not provide for exceptions from liability for normal or necessary computer dealings. For example, it may be difficult to determine whether a seller of computer software is a legitimate vendor, although the use of "disclaimers" (from civil law) in relation to the application of the provision reflecting the innocent intention of the vendors, may be possible.

To deal with the dual interpretation issue, a "knife-to-gun" approach has been proposed. Knives and guns have a dual use nature because they can either be used as tools or as lethal weapons. Unlike a knife, the owner of a gun requires a licence to carry one. In the same way, hacking tools could be classified under the "gun" state instead of the "knife" state, taking into account their potential for harm. It is submitted that hacking tools should only be possessed by an authorised person, who has some sort of "security licence" issued by a specific government body. Furthermore, to obtain such a licence, an individual must have had ethical hacker training and be certified. If the hacker is self-trained, then the granting of the licence will be determined on a case-by-case basis by the specified government body. It is further submitted that this government body has the responsibility of determining whether an article is considered a hacking tool or not, and monitoring the sale and purchase of the tools by licensed vendors. To achieve this, new legislation has to be enacted to ensure that hacking tools can only distributed by such vendors. Although these measures may be seen as a limitation on the freedom of an individual to find and download technology for individual benefit, it is submitted that freedom is not an absolute concept in that it should not be a means to cause harm to others. While this proposal may cause some inconvenience to legitimate hackers, the inconvenience caused by DoS and DDoS attacks is much more.[147]

Apart from the above issues, Section 3 the broad meaning of the offence of 'impairing' (in "with intent to impair, or with recklessness as to impairing…") also causes a problem. The issue here is that users may find it difficult to distinguish between 'impairing' caused by malicious code and programs, and that caused by their computer systems simply being out of date. Users may be uncertain about whether to lodge a complaint because they would be unsure if their systems had indeed been compromised or not. To meet this problem, it has been proposed that the government should provide subsidies to users who cannot afford to upgrade their systems. In this way, 'impairing' caused by users resorting to unreliable counterfeited resources to upgrade their systems will be overcome.[148]

According to Section 3A of Computer Misuse Act 1990 "A person is guilty of an offence if he makes, adapts, supplies or offers to supply any article intending it to be used to commit, or to assist in the commission of, an offence under Section 1 or 3".[149] The prosecutor must prove that "the defendant engaged in the relevant conduct with the intention that the article was intended to be used or to assist in the commission of the relevant offence".[150] Even if that conduct was ineffective on that act, it could still be an offence under this section due to the necessary intention of the defendant.

The second offence in Section 3A is "to supply or offer to supply any article believing that it is likely to be used to commit, or to assist in the commission of, an offence under Section 1 or 3".[151] In order to qualify for this offence, the perpetrator must be engage in a reckless

[147] R. Rahman, 'The legal measure against denial of service (DoS) attacks adopted by the united kingdom legislature: should malaysia follow suit?', *International Journal of Law and Information Technology* (2012) 20(2), pp. 89–100. Available at: http://ijlit.oxfordjournals.org/ (26 Jun 2012).

[148] R. Rahman, 'The legal measure against denial of service (DoS) attacks adopted by the united kingdom legislature: Should Malaysia follow suit?', pp. 89–100.

[149] Section 3A Computer Misuse Act 1990 (Act 3).

[150] J. Clough, *Principles of Cyber-crime*, p. 129.

[151] Section 3 Computer Misuse Act, (Art 3A (2)).

"supplying" or "offering to supply". In such a situation, the person who produces and distributes some information does not have any intention of using it, but he recognises that it may be used by others. Consequently, the prosecution must prove that an actual belief and a probability are insufficient. The third offence under this section "is to obtain any article with a view to its being supplied for use to commit, or to assist in the commission of, an offence under Section 1 or 3".[152] This is aimed at persons who obtain such articles for the purpose of providing them to others to commit an offence, since according to this Act possession is not an offence even if accompanied by the intent to commit an offence.

Supporting terrorist activity qualifies for prosecution of the activity in US according to Section 2339A and B of the US Code, even if the terrorist activity is not completed. Thus, the federal prosecutor can charge terrorists and supporters early in their conspiracies, because there is no need to link 2339B offences to specific events or attacks.

3.5.4 Unauthorised Acts with Intent to Impair

A. The UK

It was recommended by the Law Commission to establish an offence of causing an unauthorised modification to programs or data held on a computer. Therefore, this offence is included in Section 3 of the Computer Misuse Act which was amended by the Police and Justice Act 2006. The main intent of this clause is to make DoS attacks illegal as one of the tools of launching a cyber terrorist attack.[153]

Consequently, it was revamped completely through Section 36 of the Police and Justice Act 2006. The title of that section was relabelled from 'Unauthorised modification of computer materials' to 'Unauthorised acts with intent to impair, or with recklessness as to impairing the operation of a computer'. This amendment was to

---

[152] Section 3 Computer Misuse Act, (Art 3A (3)).

[153] In simple terms, the original Bill of the 2006 amendment tended to keep the old Section 3 and proposed Section 3A to focus on the problem of DoS attacks. Nevertheless, the bill redrafted the wording of Section 3 and introduced a new version of Section 3A.

fulfil international and regional European standards. This is because Article 5 of the Convention on Cybercrime uses the words 'inputting', 'transmitting', and 'altering' which obligates member states to establish criminal offences of the 'serious hindering of the functioning of the computer system'. Article 3 of the EU Council Framework Decision 2005/222/JHA uses the words 'damaging', 'deleting', 'suppressing', and 'rendering inaccessible' while referring to 'serious hindering or interruption of the functioning of an information system'.

All these words in the Convention and the Decision are symptoms of DoS attacks. Such amendments originated from the controversial decision in *R v. Lennon* where the judge decided that the Computer Misuse Act did not cover DoS attacks. The decision of the judge was based on the reasoning that each email that was sent in a DoS attack caused an authorised modification to the email server, whereas the Computer Misuse Act declares "unauthorised modification" illegal. Despite the reversing of the decision by the High Court following an appeal by the prosecution, it initiated a call by the All-Party Internet Group of MPs (APIG) to amend the Computer Misuse Act with regards to the act of unauthorised modification.

Section 36 of the Police and Justice Act with the heading "Unauthorised acts with intent to impair operation of computer" provides that:

> 1. a person is guilty of an offence if: a. he does any unauthorised act in relation to a computer b. at the time when he does the act he knows that it is unauthorised; and a person convicted of the offence maybe sentenced to a maximum of 12 months imprisonment on summary conviction or ten years on indictment.

The effect of an unauthorised act must be:

> a. to impair the operation of any computer ;b. to prevent or hinder access to any program or data held in any computer: c. to impair the operation of any such program or the reliability of any such data: or d. to enable any of the things mentioned in paragraphs (a) to (c) above to be done.

Prior to the new amendment, the term 'unauthorised modification' was used. Then it was replaced with 'any act in relation to a computer'. The new version of Section 3 applies to a wide range of 'conduct', including to DoS attacks. The impairment discussed in this section must have been intended or foreseen. Such intention is not required to be related to any particular computer, program or data, or to any particular kind of program or data. The language of subsection (c) of the Act shows that if the operation of any such program or the 'reliability' of any such data is impaired, the requisite intent existed. The reliability is affected when the computer is made to record the false information, and to show the false origination. That is, it shows that it came from one person, but it actually came from someone else.[154]

Such offences are encompassed in Articles 4 and 5 of the Convention on Cybercrime, relating to data and system interference. Data interference consists of offences with intent and without right in "damaging, deletion, deterioration, alteration or suppression of computer data".[155] The Convention authorises member parties to require the conduct to result in serious harm in order to be criminalised.[156] Article 4 of the Convention concerns intentional damage to data solely, and Article 5 relates to "system interference" which is intentional and without right, accessing and hindering the functioning of a computer. Such hindering of a computer system may occur either by modification of data or by restricting the function of the computer without any modification of data, such as in DoS attacks in cyberterrorism cases. "Therefore, these offences apply to intentionally hindering the use of a computer system by either using, or influencing computer data".[157]

The fault element is very important in this section. It is not intended to cover the offences occurring by unauthorised use but

---

[154] J. Clough, *Principles of Cyber-crime*, pp. 114–116.

[155] Article 4 Convention on Cybercrime.

[156] Convention on Cybercrime, Explanatory Report. Available at: http://conventions.coe.int/Treaty/en/Reports/Html/185.htm (26 June 2013).

[157] J. Clough, *Principles of Cyber-crime*, p. 101.

without any intention. Mere use without intention will not fall within this section.[158] This Act emphasises the point that impairment will be caused but it does not matter that the conduct is either intentional or reckless. It can be used in several situations; for instance, when a user intentionally causes the deletion of a program or data held on a computer and the question of how this conduct happened is not important. Other instances may be when the offence is committed when the data is added to a computer, or a computer is infected by a virus or the offence may be committed with a logic bomb or by adding a program to the computer system with the intent to cause inconvenience to the computer user. In fact, according to Section 3(5) of Computer Misuse Act, it is not important whether a modification is permanent or temporary. Furthermore, it does not require the degree of impairment to be either substantial or significant.

According to Section 3(3) of the Computer Misuse Act, if a person creates a computer virus and sends it out into the world with the intent of infecting other computers, the offence is committed. In the case of Christopher Pile, aka "the Black Baron", he created a virus on the internet. Every computer that downloaded this program was infected. The impairment caused by the virus was estimated at US$500,000 and he was convicted under the Act and sentenced to 18 months' imprisonment. Therefore, the Computer Misuse Act can be used against those who intentionally cause a computer to be infected. It is difficult for the prosecutor to prove the intent of the perpetrator. Notably, the government and the National High Technology Crime Unit (HTCU) believe that Section 3 covers DoS attacks.

One of the most significant issues which arises regarding unauthorised access is the DoS attack offence. This type of attack can be launched against commercial websites and is simultaneously a tool for cyber attackers to launch their cyber terrorist attacks. The UK has one of the highest proportions of "bot-infected" computers, because of the rapid take-up of broadband activity. The perpetrators of these activities mostly use computers called "zombie computers" or "botnets" that

---

[158] S. Fafinski, *Computer Misuse, Response, Regulations and the Law*, 1st Edn, Willan Publishing, United Kingdom, 2009, pp. 30–31.

act under the control of the perpetrator without the real owner's knowledge. Most cyber terrorists launch DoS attacks as seen in previously mentioned cases in this book. The Computer Misuse Act attempts to cover this kind of offense. Section 3 of the Act is able to address all such activities, and such activities have the necessary intent to cause modification or impairment. In the original Section 3, such offences were not considered under this Act. But after a case which happened in 2005, where a teenage boy carried out a DoS attack against his former employer using a specialist email bomb, Judge Grant at Wimbledon Magistrates' Court in November 2005, stated:

> … the individual email caused to be sent each caused a modification which was in each case an 'authorised' modification. Although they were sent in bulk resulting in the overwhelming of the server, the effect on the server is not a modification addressed by Section 3 of the Computer Misuse Act.

However, the sub-divisional court of appeal stated, "the owner of the computer able to receive emails would ordinarily be taken to have consented to the sending of emails to the computer. However, such implied consent was not without limits, and it plainly did not cover emails that had been sent for the purpose of communication with the owner but for the purpose of interrupting his system".

This case in which the court found that a DoS attack could be prosecuted under the Computer Misuse Act 1990 was *DPP v. David Lennon*. The defendant, David Lennon, was dismissed from the Domestic and General Group after working for them for just 3 months. Following his dismissal, he downloaded Avalanche v3.6, a mail bombing program, and sent approximately five million emails to different Domestic and General email addresses. He was first convicted under Section 3 of the Act with causing an unauthorised modification to a computer. The judge then dismissed the case against Lennon on the basis that this section was enacted for dealing with the sending of malicious code, and that it was not proper to extend it to the sending of uninfected emails. In a related case, the public prosecutor appealed and the divisional court held that the owner gave an implied consent to receive email as part of the email service's basic function. This consent had limitations that were exceeded by Lennon, making the act unauthorised.

Some of these attacks disable the communication link to the target computer and not the target itself. Therefore, the original Section 3 only covered those computers that formed part of the network, like routers. As a result, the government, in reviewing the Computer Misuse Act 1990, amended the Computer Misuse Act to address the lacuna of DoS attack activities; and furthermore, to comply with its international commitments under the Cybercrime Convention and the European Framework Decision. Consequently, it added the phrase "any point in a network to change the situation of the crime".[159]

B. Malaysia

The offence of modification is extensively defined in Section 2 of the CCA. These include:

(a) Altering or erasing any program or data held in a computer;
(b) Introducing or adding data to the contents of a computer; or
(c) Any event that impairs the normal operations of a computer, and any act that contributes towards causing such a modification.

The term unauthorised modification is extensive enough and has the potential to cover logic bombs and time-bombs beyond the modification. There must be strong evidence of unauthorised modification for conviction. In *R v. Vastal Patel* involving the development of the bespoke system, the database tables of the computer started to disappear and the wrecking program were found on the suspect's computer. The suspect was accused of erasing the tables so as to prolong this lucrative program. However, as there was no compelling evidence that he was responsible, the accused was acquitted by the jury.[160]

---

[159] S. Fafinsky, *Computer Misuse: Response, Regulation and The Law*, pp. 45–50.

[160] A. Abdul Rahim *et al.*, Cyber-Crimes: Problem and Solutions under Malaysian Law.

Section 4 of the Computer Crime Act 1997 is referred to the ulterior intent. According to this provision a person shall be guilty if he secures unauthorised access with the intention of either committing a further offence involving fraud or dishonesty or which caused injury as defined in Penal Code or facilitating the commission of such an offence, whether by himself or by other person. It is not important to commit the act at the same time when the unauthorised access is occurring. As long as the intention of the criminals is proved, he will charge under this provision. In case *R v. Thompson*, the accused devised a program which instructed the computer to transfer sums from a dormant account to an account which he had opened with the bank. In this case the ulterior intent would be committed when the accused committed the first part of the crime which would be unauthorised access with the intent to commit further offence.

Section 5 of the Computer Crime Act 1997 of Malaysia provides for the offence of unauthorised modification of the content of any computer. This provision confers a definitive punishment on those who cause the unauthorised modification of a program or data, whether the intent to make the modification was targeted at a particular computer, data, or program or not. The awareness of the offender that his act would cause unauthorised modification of the content of any computer is important. Section 5 treats as a criminal any act which knowingly places a virus on a computer as viruses cause modifications of programs or data.

Malaysia goes beyond the recommendations of international organisations such as the OECD and the Council of Europe in this matter. This is because they criminalise modification of data or programs that damage a computer system or impair its function, while the Malaysian Computer Crime Act will allow prosecution for the release of a virus, even if the virus is non-destructive.[161] In other words, according to the Malaysian Computer Crime Act, releasing a virus is punishable even if it does not cause any modifications in computer programs or data.

---

[161] D. L. Beatty, 'Malaysia Computer Crime Act 1997 gets tough on cyber-crime but fails to advance the development of cyber law', p. 360.

According to Section 5, it is immaterial whether the unauthorised access is done permanently or temporarily. The penalty can either be a maximum fine of RM100,000, a maximum prison sentence of 7 years, or both. If the modifications were effected to cause injury, then the maximum fine and imprisonment is RM150,000 and 10 years, respectively. Similarly, cyber terrorists modify the contents of computers to deploy their attack. They alter information or input data with the intent of disrupting the functioning of a computer or commercial exploitation to inflict political, religious, and economic harm.

Section 5 of Malaysia's Computer Crimes Act is quite similar to Section 3 of the UK's Computer Misuse Act 1990 where acts are considered criminal that cause unauthorised modifications to the contents of any computer regardless of whether they were directed at any particular computer, data, or program. Section 3 of the UK's Act was used to successfully convict Christopher Pile a virus programme writer known as "The Black Baron". In November 1995, Pile pleaded guilty to charges of violating Section 3 of the Computer Misuse Act by writing two viruses which eventually spread globally and was sentenced to 18 months in prison. The fact that Malaysia has generally included a provision in her laws that has contributed to the conviction of such a high-tech criminal should reassure Malaysia Smelting Corporation (MSC) investors of the commitment of the government towards protecting their interests.[162] The elements of impair can be seen in the offence of modification in Section 5 of CCA 1997 and the offence of ulterior intent in Section 4 CCA.

3.5.5 Disclosure of Information

The federal espionage law outlaws spying and bans disclosure of information potentially detrimental to US national defence. This is tracked essentially through paragraph 1030(a)(1) and enacted as part of the Act. It has been amended in order to more closely track the other espionage laws. It merged the espionage elements with

[162] D. L. Beatty, 'Malaysia Computer Crime Act 1997 gets tough on cyber-crime but fails to advance the development of cyber law', p. 362.

computer abuse. The elements that constitute this crime include: anyone who,

- Either
  (1) Wilfully disclosing;
  (2) Wilfully attempting to disclose; or
  (3) Wilfully failing to return.
- Classified information concerning national defence, foreign relations, or atomic energy
- With reason to believe that the information either
  (1) Could be used to injure the US; or
  (2) Could be used to the advantage of a foreign nation.
- When the information was acquired by unauthorised computer access.

The intent element is pegged high in paragraph 1030(a)(1). Thus, these criteria must be fulfilled by the offender: (1) purposefully transmits or retains information that (2) he has reason to believe could be used to the injure the US or benefit another country, and (3) that he has obtained through access to a computer that he knows he had no authority to access. Moreover, according to the Supreme Court statement in *Gorin v. U.S*[163] these requirements must be acted on in bad faith. Such crimes, according to these guidelines, are punishable by imprisonment for not more than 10 years (not more than 20 years for second and subsequent offences) and/or a fine of not more than US$250,000 (not more than US$500,000 for organisations).[164]

The crimes of attempt, conspiracy, and complicity are the same as the other crimes under paragraph 1030(a) and subject to the same penalties. Put another way, those who attempt or aid and abet the violation of another are subject to the same penalties as those who

[163] 111 F.2d [1940], p. 712.
[164] 18 U.S.C. 1030(c)(1), 3571.

commit the substantive offence.[165] According to general espionage sentencing guidelines, a base sentencing (level of 30) is applied for the violation of paragraph (a)(1) which begins at 8 years' imprisonment, and if the top-secret (level 35) is involved, it begins at 14 years. The federal crimes of terrorism are included in paragraph 1030(a)(1), and the minimum violation attracting sentencing for terrorist purposes is located at level 32.[166]

Studying 1030(a)(1) and (5), it appears that the drafters of the federal terrorism crimes intended to include the attempt of a terrorism crime as a complete crime. It implies that they accepted intent as the base element of cyberterrorism and they ignored the result of the attempt.

Section 1030(a)(7) of the US Code also prosecutes extortionist hacking that prohibits "any communication containing any threat to cause damage to a protected computer" with the "intent to extort from any person any money or other thing of value" in interstate or foreign commerce which can be applicable to cyberterrorism cases as a means of attack.[167] It was enacted in 1996 out of concern that the protected property under this law does not include operation of a computer, the data programs stored in a computer and decoding keys of encrypted data. Thus, in 2008, with the recommendation of the Department of Justice, the following parameters were added to the section to "cover the situation in which a criminal has already stolen the information and threatens to disclose it unless paid off" and in which "other criminals cause damage first — such as by accessing a corporate computer without authority and encrypting critical data — and then threaten that they will not correct the problem unless the victim pays".[168]

This statute does not specify any threshold value and it depends on dual jurisdiction in order to qualify for a successful prosecution. For e.g. the threat has been transmitted in interstate or foreign commerce,

---

[165] 18 U.S.C. 1030(b).

[166] U.S.S.G. Section 3A1.4.

[167] 18 U.S.C 1030 (a)(7).

[168] Charles Doyle, Cyber-crime: An Overview of the Federal Computer Fraud and Abuse Statute and Related Federal Criminal Laws, p. 62–63.

and the transmitted threat must be directed against a protected computer which is used by or for the federal government or which affects interstate or foreign commerce. Nevertheless, it is not necessary for the government to show that the defendant knew that the threat had been transmitted in interstate commerce. It covers any interstate or international transmission of threats against computers or computer networks in any way by mail, telephone calls, electronic mail, and computerised messages. Section 1030(a)(7)(B) and (C) address different threats. Subparagraph (B) addresses threats to breach the confidentiality of data and subparagraph (C) addresses "threats to fail to undo damage already inflicted if extortionate demands are not met".[169] Section 1030(a)(7) is often applicable in conjunction with Section 1030(3) which is the access to use, or the ability to use, a government computer. It requires that a user access intentionally a government computer exclusively, but it does not require any damage to be alleged.

One of the main and significant differences between the US' anti-terror law and that of Commonwealth countries such as the UK is that, it makes a distinction between acts causing death and those causing lesser levels of harm. Other countries and the UK provide the maximum penalty of life imprisonment for all politically motivated activities that interfere with electronic systems and infrastructure, and do not pay attention to the intention of an offender as to whether he/she tends to cause a greater level of harm, injury, or death. The US calibrates its anti-terror law's penalties according to the level of harm caused by cyber attacks and keeps the maximum penalty only for cyber attacks that cause or attempt to cause death. The US and the UK have enacted laws providing sweeping terrorism penalties which promise maximum penalties for any acts of terrorism directed at electronic systems and other infrastructure. The words "any acts" include cyberterrorism as one of the terrorism activities.

Taken together, it may be concluded that compared to other jurisdictions (e.g. the UK and Malaysia and even other countries) the US has done a lot more work on its anti-terror law in order to provide

---

[169] Charles Doyle, Cyber-crime: An Overview of the Federal Computer Fraud and Abuse Statute and Related Federal Criminal Laws, p. 64.

a precise law which covers every possible case which may occur. It also makes a searching distinction between different levels of harm. However, it appears that all these three jurisdictions provide extraordinary low harm requirements for acts of terrorism directed against electronic systems and other infrastructure. It is different in other developed countries; Canada requires that the targeted infrastructure be "essential" to the country, and New Zealand requires the attack be "likely to endanger human life", in order to attract the maximum penalty of life imprisonment.[170] Most countries do not distinguish between the different techniques used to launch cyber attacks, so long as the attack satisfies the more general harm and fault requirement of a terrorist act. However, all of them do distinguish a cyber attack directed against electronic systems and infrastructure.

In general, the prosecution's involvement in investigating cyber crime varies in each jurisdiction. In civil law countries, the investigation process will begin with the prosecutor's order, while in the US the method of investigation is totally different. There, the criminal charge plays an eminent role. In some countries the prosecutors have an active role in legislation. In fact, if the prosecutors are involved in the investigation from an early stage, they can procure the proper evidence to ensure efficiency.

In common law countries, prosecutors must decide what cases they take on, while in other legal systems, prosecutors are required to prosecute when sufficient evidence is available. These decisions are reached by the prosecution policies in some modern jurisdictions based upon the seriousness of the offence and the sufficiency of the evidence, but in others, the law dictates it. The search warrant is applied in two ways in different countries. In some of them, the law enforcement officers must obtain judicial authority before searching and seizing, while in others this authority is vested in prosecutors or police. Countries that vest the search warrant in the police or prosecutor give a large amount of authority to its police, which may be abused by them. Prosecutors in common law jurisdictions must

---

[170] K. Hardy, 'WWWMDs: Cyber-Attacks against infrastructure in domestic Anti-Terror Law', *Computer Law and Security Review* (2011) 27, p. 153.

present the evidence in court; in other jurisdictions they just assist the judge in fact finding.[171] Put simply, in common law countries like the UK the prosecutors decide what cases they take on. While, according to the US' Fourth Amendment of the Constitution, in order to obtain a criminal warrant, a federal agent must prove that under the circumstances known to him or her there is a reasonable belief that a person has committed, is committing, or is about to commit a crime.

In general most nations have enacted specific computer crime provisions aimed at countering cyber attacks by including a wide array of activities that could be designated as cyber crimes and cyberterrorism. Countries utilise their computer crime and anti-terrorists acts to check cyber terrorists and have met with some success in these areas. The following section will consider the application of these statutes in Estonia which has experienced what is considered a landmark case of cyber terrorist attack.

3.5.6 Virtual Weaponry Used By Terrorist

The domain of cyber space is obviously not as physically transparent as the domains of air and space. Cyber space attacks, just like criminal hacks, are designed to affect electromagnetic data in various ways, which then impact the adversary's cyber space to create an operational advantage for the attacker. Thus, the basic legal framework of the Cybercrime Convention is beneficial in the context of military operations in cyber space. Although that Cyber Crime Convention does not provide all key terms, it points to the criminal codes of member states. The US as a larger user of cyber space is a leading nation in prosecuting these crimes. Therefore, so many nations follow the US in solving cyber space crimes. Section 1030 of the US Code provides a clear, judicially accepted definition of damage in cyber space. It defines it as "any impairment to the integrity or availability of data,

---

[171] S. Brenner *et al.*, 'Transnational evidence gathering and local prosecution of cybercrime', *John Marshall Journal of Computer and Information Law* (2002) 20, p. 347.

a program, a system, or information". By borrowing from this definition, it is possible to intelligently and sensibly develop a definition of the word "cyber space weapon". From the above, the definition of weapon can be said in its most simple terms as "something that causes damage". According to the definition of damage from Section 1030, Todd proposed the following definition for cyber space weapon: "Any capability, device, or combination of capabilities and techniques, which if used for its intended purpose, is likely to impair the integrity or availability of data, a program, or information located on a computer or information processing system".

In order to determine when 'armed attack' is constituted under international law, the way a cyber space weapon is used is very important. The intent of the user in the main issue in criminal law. Proving the intent of the state actor in 'armed attack' is very difficult specially when just relying on evidence. State actors have a responsibility, as sovereigns with power and control, to conduct their affairs with due regard for their international legal obligations when they have knowledge of activities within their legal or physical boundaries. "Cyberspace attack occurs when a state knowingly uses or knowingly acquiesces to an entity under its legal control or within its territory using a cyberspace weapon against the people or property of another state".[172]

According to this definition additional responsibility is imposed on states under international law when a state knowingly allows a person or entity to use a cyber space weapon against the people or property of another state. The cyber space attack's definition is not intended to hold states responsible for the nefarious efforts of criminals, unless the state is aware of the criminal's efforts and knowingly acquiesces to the use of a cyber space weapon. International organisations place a higher burden on states in order to encourage state to refrain from acquiescing to certain acts committed from their territory against other states. For instance, according to the UN Security Council Resolution 1373 which is decided under Chapter VII of the

[172] H. Todd, 'Cyber attck in cyberspace; Dettering asymmetric warfare with an asymmetric definition', *Air Force Law Review* (2009) 64, p. 85.

UN Charter, all states shall take the necessary steps to prevent terrorist acts, to prevent terrorists from using their territories for terrorist purposes, and to provide early warning to other states. Although not all of them, but some cyber space attacks could fit within the resolutions' already accepted prohibitions related to harbouring terrorists, preventing use of territory by terrorists, and acquiescing to those who cause civil strife. The definition of armed attack in cyber space really only increases the degree of state responsibility for non-state actors by a small amount. According to the cyber space attack definition which is proposed by Todd a clear responsibility would be placed on all states. It requires a reasonable *mens rea* for a host state for acknowledging the impending or ongoing use of a cyber space weapon.

3.6 ESTONIAN LEGAL RESPONSES TO CYBER ATTACKS: A CASE STUDY

Estonia was the location of a land mark case in the cyberterrorism field. Following the occurrence of the attacks against Estonia, it was decided that additional protective measures should be taken. Based on the attacks and communications of the European Commission and its regulating policy to fight against cyber crime in the EU, the Estonian government began to analyse regulations on cyber crime and related legal fields in national as well as international law. At this time, Estonia has yet to establish suitable legal terms to respond to such attacks. The investigations in Estonia were based on national criminal law, since they did not have any specific provisions in their Penal Code regarding computer crime. Therefore, they followed the rule of criminal proceedings.[173] Estonia has tried to obtain the cooperation of several countries. The request for cooperation with Russia remains unanswered, since it has not criminalised computer crime. Thus, there can be no prosecution without investigation. In fact, cross-border criminal proceedings are not possible and even if they

[173] M. Tikk *et al.*, Legal and policy evaluation: International coordination of prosecution and prevention of cyber terrorism, In *Centre of Excellence Defence Against Terrorism. Response to Cyber Terrorism*, IOS Press, Turkey, p. 101.

were possible, it would be difficult to obtain a prosecution as there is no general agreement on what data the communication service providers need to gather or submit to authorities. Normally, a person must have direct involvement in the commission of the crime and evidence must be shown in order to convict such a person.

The sources of Estonian criminal procedural law are based on the Constitution of Estonia, the generally recognised principles and provisions of international law, and international agreements binding on Estonia. Thus, "the Code of Criminal Procedure and other legislation which provides for criminal procedure; and decisions of the Supreme Court in issues which are not regulated by other sources of criminal procedural law but which arise in the application of law. Therefore, any person subject to proceedings has to take into account the legal effects of the Estonian law as well as the treaties concluded with other countries to the extent that is predictable according to the relevant agreements".[174]

Therefore, because of the lack of law enforcement, Estonian national law has been reviewed on a much broader plane. It thereby provided for a more efficient set of investigation and prosecution tools to deal with computer-related crimes which can potentially be committed by terrorists. Estonia has revised its whole legal framework as a consequence of the attacks. The amendments were implemented in IT legal policy and legal regulation of information systems, rights and obligations of internet, and other communication service providers in order to understand what units of data needed to be logged and saved, and to determine if this information would be available for investigation.[175]

3.6.1 Legal Development in Estonia after the Attack

The 2007 attacks gave rise to major changes in Estonian legislation (and also sped up the development of changes already in progress).

[174] E. Tikk *et al.*, Legal Cooperation to Investigate Cyber Incidents: Estonian Case Study and Lessons, pp. 288–295.

[175] M. Tikk *et al.*, *Response to Cyber Terrorism*, p. 101.

There were legal amendments in areas of law relating to cyber security, such as criminal law (including parts of criminal procedure) and crisis management law. Other laws such as the Electronic Communications Act 2000 were also updated but did not involve considerable changes in the area of cyber security. However, since the attacks were treated by national authorities as acts of crime, the Estonian attacks did not directly impact legislation on armed conflict.

Cyber crime-related provisions in Estonia's Penal Code were reviewed in order to be harmonised with the Council of Europe's Convention on Cybercrime (Council of Europe, 2001) and the Council Framework Decision 2005/222/JHA on attacks against information systems (Council of Europe, 2005). The amendments to the Penal Code included provisions addressing attacks against computer systems and data. "It widened the scope of specific computer crime provisions, added a new offence of the preparation of cyber crimes, modified the provision concerning acts of terrorism, and filled an important gap in the Penal Code by enabling differentiation between cyber attacks against critical infrastructure and ordinary computer crime".[176]

The amendments to the Penal Code sought to address the existing regulatory limitations in the application of the Code of Criminal Procedure (CCP) to the attacks.[177] Sections 110–112 of the CCP stated that evidence could only be collected by surveillance activities in a criminal proceeding if the collection of evidence was, (a) precluded or very complicated, and (b) the criminal offence under investigation was, at least, an intentionally committed crime for which the punishment was at least 3 years' imprisonment.[178] But none of the offences committed during the Estonian attacks met the requirement of

[176] C. Czosseck, Estonia after the 2007 Cyber attacks: Legal, Strategic and Organizational Changes in Cyber Security, *Proceedings of the European Conference on Information Warfare and Security*, Cooperative Cyber Defence Centre of Excellence, Tallinn, Estonia, 2011, p. 60.

[177] Estonian Code of Criminal Procedure, (2010b), Available at: http://www.legaltext.ee/text/en/X60027K6 .htm. (25 Jul 2012).

[178] Estonian Code of Criminal Procedure, (2010b), Available at: http://www.legaltext.ee/text/en/X60027K6.htm. (25 Jul 2012).

"three years" imprisonment and this prevented the use of surveillance measures.[179] Thus, changes to the Penal Code now prescribe higher maximum punishments and also address corporate liability for cyber crime offences.

A new act which was enacted as a result of the attacks is the Emergency Act which was adopted in June 2009. The Act reviewed the framework of national emergency preparedness and emergency management, including available responses to cyber threats. It then developed a comprehensive approach via measures such as preventing, preparing for, and responding to emergencies, and mitigating the consequences of emergencies.[180] Public service providers and information infrastructure owners have been given the responsibility of daily emergency prevention and for ensuring stable levels of service continuity.

Providers of vital services are required to prepare and present a continuous operational risk assessment[181] and an operation plan[182] to notify citizens of major disturbances in service continuity, as well as providing necessary information to supervisory bodies. Also, there are specific provisions that address threats against information systems. One of these is the provision that providers of vital services are obliged to guarantee the smooth application of security measures in information systems and assets used in providing vital services.

An analysis of Estonian legislation relating to an information society shows that security for an information society needs to be comprehensively supported by standards covering several legal disciplines. The Estonian legal framework's broad approach to cyber incidents

---

[179] Estonian Government, (2007b), Explanatory Memorandum to the Draft Act on the Amendment of the Penal Code (116 SE) (In Estonian). Available at: http://www.riigikogu.ee/?page=pub_file&op=emsplain&content_type=application/msword&u=20090902161440&file_id=198499&file_name=KarSseletuskir(167).doc&file_sise=66048&mnsensk=166+SE&etapp=03.12.2007&fd=29.10.2008. (25 July 2012).

[180] Ministry of the Interior, Department of Crisis Management and Rescue Policy (in Estonian). Available at: http://www.siseministeerium.ee/elutahtsad-valdkonnad-ja-teenused-2. (25 Jul 2012).

[181] Section 38 Emergency Act, June 2009.

[182] Section 38 Emergency Act, June 2009.

brings together the areas of private and public law, and includes criminal law, crisis management regulation, and wartime law/national defence legal order.

Thus, the Estonian incident has showed the importance of countries realising that international cyber security regulation must involve a wide range of legal areas. It also raised the need for a review of relevant regulatory frameworks to identify potential "grey areas". The central issue among national legal systems appears to be a review of criminal law. This is because it is domestic criminal law (and related provisions for investigation and prosecution) that should regulate such attacks against critical infrastructure, politically motivated cyber attacks and cyberterrorism. It is crucial that comprehensive national implementation of the Council of Europe's Convention on Cybercrime be carried out, especially in regard to the cross-border nature of cyber crimes.

3.6.2 Organisational Development in Estonia: Post Attack

Following the attacks, Estonia has made some key organisational changes to better deal with cyber threats. Before the attack, organisations which were dedicated to cyber defence were few. The most important organisation was the Cyber Security Council under the Government Security Committee, whose formation was recommended by the National Cyber Security Strategy. The Council coordinates inter-agency and international cyber incident response and reports directly to the Government Security Committee.

There is also the Department of Estonian Informatics Centre (EIC) which is responsible for managing and developing public information services and systems, and providing cyber security for services and systems as a state agency. Community Emergency Response Teams (CERT) was established in 2006 as a part of the EIC. The Department of Critical Information Infrastructure Protection (CIIP) was also added to the EIC in 2009. The task of CIIP is to supervise the risk analyses of critical information infrastructures and develop protective measures.[183]

[183] P. Cunha *et al.*, Evaluation of the Armed Forces Websites of the European Countries, *Proceedings of the 10th European Conference on Information Warfare and Security*, Tallin, Estonia, 2011, p. 60.

Another organisation is the Cyber Defence League (CDL) which was formed in 2009. It was founded by both the CERT organisation and informal networks of volunteers who assist CERT and cooperate against criminally motivated cyber attacks targeting critical infrastructure. The Defence League is a volunteer national defence organisation in the military chain of command. The CDL is a part of the Defence League and serves to unite cyber security specialists who volunteer their time and skills to protect the high-tech way of life in Estonia, especially by assisting in the defence of critical information infrastructures. Importantly, this is a defensive organisation, not one designed to harm political opponents via anonymous cyber attack campaigns. In January 2011, the CDL was reorganised into the Cyber Defence Unit of the Defence League, but the CDL name is still widely used.

To sum up, the Estonian event emphasised the necessity of establishing common security standards for all computer users, information systems, and owners of critical infrastructure.[184] By 2011, some progress had been achieved in the establishing of such standards for service providers, under the ambit of the Electronic Communications Act. However, more detailed rules for end users' conduct and/or legal obligations are still needed. The 2007 cyber attacks on Estonia also served as a wake-up call to nation states on the new threats emerging from cyber space. In the aftermath of the incident, Estonia developed national policies and strategies in order to counter future occurrences of cyber attacks. It developed a more comprehensive legislative landscape in several areas of law related to cyber security, criminal law, crisis management law, and national defence law.

3.7 CONCLUSION

Cyberterrorism is a consequence of new technologies. It is a kind of cyber crime produced from the convergence of cyber space and terrorism. Countries use and sometimes update their existing laws to

[184] C. Czosseck, Estonia after the 2007 Cyber-attacks: Legal, Strategic and Organizational Changes in Cyber Security, *Proceedings of the 10th European Conference on Information Warfare and Security* Cooperative Cyber Defence Centre of Excellence, Tallinn, Estonia, 2011, p. 62.

combat cyberterrorism. Pioneering countries in this field such as the US use their federal laws to combat the misuse of the internet for criminal and terrorist purposes. The US does not seek to address every action related to cyber crime and cyberterrorism, but rather focuses more on actions and policies that commonly happen. It has enacted criminal laws on cyber crime and cyberterrorism that target computers and computer networks. Countries such as the US have enacted specific criminal offenses related to ancillary cyber activities and terrorist activities that are applicable to cyberterrorism. Under US law, the same conduct may end up with multiple charges, and each element alleges a different charge. Cyber crime offenses also receive different charges according to the intent of the perpetrator. In the UK, specific actions are criminalised rather than the medium through which the actions are committed. These terrorist-related actions can take placed in cyber space and other spaces. Their illegality is not limited to their taking place in cyber space. The UK provides extraterritorial jurisdiction in the Terrorism Act of 2006; anyone committing an offense outside the UK is still guilty within. The UK contributes their Computer Misuse Act 1990 to cyberterrorism in case it would happen. Malaysia attempts to enact new law such as Security Offence Act to combat such crimes in a more proper way. Previous legislations in Malaysia regarding to terrorist acts suffer from ambiguity. In conclusion, the US in comparison with two other jurisdictions provide more comprehensive legislation for the act of cyberterrorism.

From the discussion of application of law there are some provisions that may be invoked in combating terrorism. This further legal action may be taken against the terrorists which demands specific investigation and prosecution by the enforcement. This issue will be discussed further in Chapter IV.

CHAPTER IV

ISSUES OF ENFORCEMENT IN CYBERTERRORISM

4.1 INTRODUCTION

The progress and development in computer technology have provided new opportunities for criminals, resulting in the commission of crimes which are of a different nature. Consequentially, this scenario poses new challenges to the legal systems as well as law enforcement.[1] Indeed, there are substantial differences between cyber crime and conventional crime in terms of enforcement. In cyber crimes, there are obvious difficulties in tracking, arresting, and prosecuting cyber criminals. The same can be said of cyberterrorism. These crimes can be committed from faraway places because they do not require the physical presence of the attacker. Cyber criminals can terrorise from afar by setting up, running, and executing a program at will like a time-bomb.[2] As such, commission of such crimes presents new enforcement challenges, requiring different approaches in investigation and

[1] S. W. Brenner, 'Cyber-crime investigation and prosecution: The role of penal and procedural law', *Murdoch University Electronic Journal Law* (2001) 8(1), p. 1.
[2] N. Nykodym *et al.*, 'Criminal profiling and insider cyber-crime', *Digital Investigation* (2005) 2(4), pp. 261–267. Available at: http://www.sciencedirect.com/science/article/pii/S1742287605000915 (12 April 2012).

prosecution.[3] The attackers make the task of law enforcers harder by acting as an organised group, sharing their information, and hiding their identities.[4] The attackers have several open choices in stealing data and information. They can steal from the main server, backup server, web pages, or during the transition of data between two points.[5] As the threat of cyberterrorism looms ahead and has become a global problem, it is important to look into issues of its enforcement. The issues of enforcement are divided into three parts: investigation process, prosecution process, and extradition, which is part of prosecution process.

The fact is pursuing cyber terrorists is not possible unless investigators and prosecutors are equipped with the necessary legal tools. First and foremost, there should be an all-encompassing legal definition of 'cyberterrorism', covering all forms of cyberterrorism. Secondly, there should also be adequate procedural rules for evidence gathering and investigation. Last but not least, adequate procedural rules should also be made available to prosecute cyber attackers. In a nutshell, it is extremely important for lawmakers to take such measures and come up with adequate penal and procedural laws due to the transnational nature of cyberterrorism. Such effort is needed to deny cyber terrorists any chance of taking advantage of gaps in existing laws.

This chapter aims at discussing several enforcement issues in cyberterrorism. In doing so, several issues of investigation and prosecution in cyberterrorism will be dissected. Being a unique form of crime, it is indeed pertinent to look, first and foremost, at the overview of cyberterrorism enforcement. Upon comprehending the overall enforcement scenario, the chapter will then proceed to discuss various issues and challenges in the investigation of cyberterrorism.

---

[3] S. W. Brenner, 'At light speed: Attribution and response to cybercrime/terrorism / warfare', *The Journal of Criminal Law and Criminology* (2007) 97(2), p. 400.

[4] N. Nykodym *et al.*, 'Criminal profiling and insider cyber-crime', pp. 261–267.

[5] S. Hinduja, 'Computer crime investigation in the United States: Leveraging knowledge from the past to address the future', *International Journal of Cyber Criminology* (2007) 1(1), p. 56.

Next, this chapter will focus on relevant issues and challenges on the prosecution of cyberterrorism.

4.2 CYBERTERRORISM ENFORCEMENT: AN OVERVIEW

One of the main steps in cyberterrorism incidents is the identification of the attacker. This plays an important role in ascertaining the nature of an attack, which leads to the formulation of a response to the attack. In some cases, a real-world terrorist attack may arise through activities in the virtual world. In these cases, if it is possible to match identification information between the cyber activities and an individual, then the cyber activities may be used as evidence to show intent, knowledge, and culpability. However, this kind of attribution may be difficult to achieve.[6] The enforcement process is divided into three sections: investigation, prosecution, and extradition.

The prosecution of perpetrators of online offences requires evidence. The evidence is revealed by the investigation process. Firstly, that a crime was committed and secondly, of the identity of the perpetrator. One way to prove the identity of such an offender is to obtain the identity of subscribers (e.g. name, address, telephone number) through the records of owners/Internet Service Provider (ISPs) located in the real world. This information can then be obtained through subpoenas to the owners of the virtual sites or to the relevant ISPs. This disclosure of information will be permissible because users who access the virtual world would generally sign some sort of contract allowing the owners/ISPs to provide information about their users to the government under certain circumstances, such as in a lawful investigation. In addition, the government can also require that one of the items in the contract between the owners and users will enable the company/ISP to monitor and

[6] C. L. Glaser, Deterrence of Cyber-attacks and U.S. National Security (2011) 3 George Washington Univ. Cyber Security Policy & Research Inst. Report GW-CSPRI-2011-5, 2011. Available at: http://www.cspri.seas.gwu.edu/seminar 2010_2011.html (3 Jul 2012).

control illegal conduct by its users, such as the reporting of illegal content which includes terrorist activity online. If such is the case, then the owner of the virtual site that is being used, for e.g. as a meeting point for virtual terrorist cells, is also required to report this type of conduct.[7]

The attribution of an attack for the crime or terrorism is problematic. Often, terrorists acting on behalf of a terrorist group do not take responsibility for the attack. Identifying the origin of the attacker is not an easy job. This is because online attackers use "stepping stones" which are owned by innocent parties. Therefore, while an attacking server might be in China, it might mean that the attack originates from China, or it might mean that it comes from another place. In physical crime and even terrorism, often the origin of the point of attack and the point of attack occurrence are closely related to each other and even if there is distance, it is a short distance. Such identification advances the process of identifying the attacker.

4.3 CURRENT INVESTIGATION PROCESS IN CYBERTERRORISM CASES

First, let us examine the current investigation process in cyberterrorism cases.

4.3.1 Current Cyber Attack Methods and the Threat they Pose

Generally, current cyber attacks comprise primarily virus and worm attacks, denial of service (DoS), web defacement of information sites and unauthorised intrusion into systems. Web defacement can be done in a second, since it is done through an exploitation tool that searches for vulnerabilities in web servers and makes a list of those vulnerabilities, so that the attacker can then choose his desired web

[7] L. Ugelow *et al.*, 'Fighting on a new battlefield armed with old laws: How to monitor terrorism in the virtual world', *University of Pennsylvania Journal of Constitutional Law* (2012) 14, pp. 1047–1048.

site and desired changes, and the program does the rest. Most of the time, the design and maintenance of web pages are done by people with limited IT skills because setting up a web page can be done easily through tools such as the Microsoft Web page design tool. A very dangerous defacement can occur when the web page is changed in such a way as to pretend to be part of someone else's system, and then an innocent user submits his or her information. Such an attack, using web pages of large companies or banks, through the use of a popup browser window that can take a client's login name and password, is called phishing.[8]

Viruses and worms are the other mediums of attack that infect computers. The difference between a virus and a worm is blurred. They infect a computer by being copied and performing a programmed function. The distinction is that a virus infects another program, while worms are self-replicating and do not need to infect another application.[9]

DoS and distributed denial of services (DDoSs) attacks have become a serious problem in the current decade. They cause interruptions to services and create irregular false signals in feedback control mechanisms that lead to significant threats to internet services.[10] Permanent elimination of DoS attacks cannot be attained yet, despite the universal transformation of the internet.[11] Thus, conceiving a mechanism to mitigate the impact of DoS attacks is the best avenue so far. One of the best methods to detect a DoS attack is by monitoring the CPU. A common way to identify a DoS attack is the high number of packets and high utilisation rate of the CPU which happens during an attack.

---

[8] L. Janczewski *et al.*, *Managerial Guide for Handling Cyber Terrorism and Information Warfare*, IDEA Group Publishing, United States, pp. 99–103.

[9] J. Clough, *Principles of Cyber-crime*, 1st Edn, Cambridge University Press, United Kingdom, 2010, p. 33.

[10] Yajuan Tang *et al.*, Protecting Internet Services from Low-Rate DoS Attacks. In: E. Goetz and S. Shenoi (Eds.), *Critical Infrastructure Protection*, Springer, Boston, 2008, pp. 251–265, 263.

[11] Wu Zhijun *et al.*, 'An approach of defending against DDoS attack', *Journal of Electronic* (2006) 23(1), pp. 148–153.

The severity of a DoS attack is captured by monitoring the CPU utilisation rate and comparing it with a previous CPU utilisation base line. When a system is targeted by a DoS attack, the service speed of the system slows down and suddenly the number of spam emails increases dramatically. Sometimes, it makes the site unavailable and impossible to access. There are many types of DoS attacks, which are either so disruptive that they prevent users from using the network services, or degrade the quality of service so as to slow down the service. Detecting a DoS attack is a very difficult and confusing job because the detector must distinguish between genuine and bogus data packets in order to determine the organisation that launches an attack. However, no perfect technique for this has been developed so far. To investigate a DoS attack, the first step is to identify the DNS to trace the Internet Protocol (IP) address by using DNS logs, and then the investigator can identify the various attacks originating from that attacker.[12]

4.3.2 Conducting Investigation and Tracking Cyberterrorism: Current Method

As mentioned above, it is better to include cyberterrorism crimes as a part of computer crime. Although there is no agreed definition of computer crime among various countries, computer crime mainly refers to a limited set of offences which are defined in the computer crime law of the US, the UK, and Malaysia in this book. The primary goal of investigation is to uncover and discover the truth. The investigation process is based on digital evidence. The process of investigation is the same as for conventional crime. Digital investigation depends on several factors, such as the type of communication device or the field of investigation i.e. if criminal or military. Despite the variations that exist in the forms of investigation, there are many similarities observed among the investigation methods.

[12] EC-Council, *Computer Forensics Investigating Network Intrusion and Cyber-Crime*, EC_Council Press, United States of America, 2010, pp. 5–9.

A. Collecting, preserving, analysing, and presenting digital evidence

The main distinction between cyber crime and other crime is the nature of evidence. In the forms it takes, how it is stored, where it is located, how it is found, and what it will tell you. Digital evidence continues to pose substantial challenges, since it is intangible, but it may also be massive in quantity and often it is volatile. However, the basic principle remains the same in digital evidence as in physical evidence. The examination of the data must entail no alteration or modification of data. The prosecutor would preserve and copy the data; then all the examinations would be done on the copy. The necessary changes in the original data should be documented and justified; otherwise, they can minimise defence challenges to the integrity of data. The evidence which is presented in a court must remain in a manner that does not change its meaning. In the digital evidence, three basic issues will always arise: first, the identity of the author of the evidence may be questioned by the defence. For instance, they may claim that it was not that person; it was somebody else. Second, the defence may claim that the original evidence was tampered with. Third, he/she may claim that the unreliability of computer program caused inaccuracies in the output.

i. *Collecting digital evidence*

The first step in investigation of cyberterrorism cases, after getting permission to investigate through a court judgment or mutual legal assistance, is to track the source of the computer intrusion to identify compromised computers. The process involves tracing back from the targeted computer, through intermediary computers, to the computer from which the attack originated. It needs the assistance of foreign law enforcement agencies and agencies connected to the internet because cyber attacks can occur via multiple ISPs located in various countries. Due to the volatile nature of digital data, the prosecutor must take provisional measures in each relevant State to preserve traffic of data and allow digital data to be available for investigation and prosecution.[13] When electronic evidence is stored by an ISP on a computer located in Country A, law enforcement must seek assistance from law enforcement authorities in Country A. Law enforcement officers exercise their functions in the territory of another country only with the consent of that country.

Therefore, law enforcement should only make direct contact with an ISP located in Country A with (1) prior permission of the foreign government; (2) approval Office of International Affairs (OIA) or (3) other clear indicia that such practice would not be objectionable in Country A. In the case that Country A cannot otherwise provide informal assistance, requests for evidence usually will be made under existing Mutual Legal Assistance Treaties (MLATs) or Mutual Legal Assistance Agreements, or through the Letters Rogatory process. The requests for assistance should be made officially by the OIA to the designated "Central Authority" of Country A or, in the absence of an MLAT, to other appropriate authorities that usually located within the Ministry of Justice or other ministry in Country A.

The OIA has lawyers responsible for every country and region of the world. Law enforcement should contact the OIA as soon as a request for international legal assistance becomes a possibility because official requests of this nature require specific documents and procedures that can take some time to process. When the law enforcement body of the requesting country has sufficient grounds to believe that electronic evidence can be found on a computer or a computer network located abroad, a request for foreign law enforcement to preserve the evidence should be made immediately. The level of success met by a request will vary relative to several factors. Such factors include whether the target country has a data preservation law and whether the requesting country has considerable contacts in law in the target country to ensure that the request will receive prompt attention. In the same context, the European Convention on Cybercrime (Convention on Cybercrime) requires all parties to be capable of effecting cross-border preservation requests; the availability

---

[13] D. Chaikin, 'Network investigations of cyber-attacks: The limits of digital evidence', *Crime Law Social Change* (2006) 46, p. 246.

of such an important form of assistance is expected to increase significantly in the near future.[14]

ii. *Preserving digital evidence*

Following the identification of the digital evidence and computer evidence, the prosecutor must take active steps to preserve such evidence since most digital evidence has a short life span.[15] They must preserve the traffic which was critical to the cyber attack and data which may reveal the source. The prosecutor will ask each ISP to preserve traffic data and disclose sufficient information in order to identify the next path of communications.[16] As soon as the source of the attack is identified and a suspect tied to a specific computer, the key terms of the investigation are found. However, it is very difficult to link a suspect to a computer, even in sabotage cases, in circumstances where the source of the attack is traced to a specific workstation, as the owner of that workstation may deny using that computer. They may claim that someone else logged on to the alleged computer and entered malicious code. The investigator has to rely on circumstantial evidence such as witness statements and digital records to place the suspect at a particular computer at a particular time. Such Digital records which may aid investigators to prove that a particular computer system was used by the suspect, may include video surveillance cameras, room access control systems, and telephone records.

The collection and transmission of digital evidence in foreign countries is possible through treaties or mutual legal assistance. The traditional method of cooperation is slow and very bureaucratic and may last several months and increase the risk of losing the criminals. The most significant treaty in this area is the Council of Europe (CoE) Convention on Cybercrime, especially Articles 20 and 21

[14] H. M. Jarret *et al.*, *Seizing Computers and Obtaining Electronic Evidence in Criminal Investigations*, 3rd Edn, Office of Legal Education Executive Office for United States Attorneys Publication, pp. 56–58.

[15] D. Chaikin, 'Network investigations of cyber-attacks: The limits of digital evidence', *Crime Law Social Change* (2006) 46, p. 252.

[16] D. Chaikin, 'Network investigations of cyber-attacks: The limits of digital evidence', p. 246.

which facilitate collection of computer data by requiring parties to introduce legislative measures "to ensure the real time collection or recording of traffic data associated with specified communications transmitted in its territory, through the application of technical means on the territory".[17] However, this cannot overcome all obstacles in law enforcement. For example, if an ISP is required to intercept data and subscriber information, this requires authorisation under national law in the form of a search warrant that will be discussed in the last section of this chapter.

In fact, cyberterrorism investigation needs to be a fast process, otherwise, the digital document may perish over time. Countries should create agents to bypass the bureaucratic procedure of treasury protocols to prevent digital documents from perishing. It would be better to establish agents to collect information from a wide variety of law enforcement agencies in a way that suits a particular investigation with a fast procedure to prevent delays in cyberterrorism investigations.

In general, the prosecution's involvement in cyber crime varies in each jurisdiction. In civil law countries, the investigation process will begin with the prosecutor's order, while in the US the method of investigation is totally different. There, the criminal charge plays an eminent role. In some countries the prosecutors have an active role in legislation. In fact, if the prosecutors are involved in the investigation from an early stage, they can procure the proper evidence to ensure efficiency.

4.4 CYBERTERRORISM INVESTIGATION IN INTERNATIONAL CONVENTIONS

The commission of traditional crime on networks causes computer and network problems both in criminal law and in investigation, prosecution, and prevention of such crimes. Such problems stem from a variety of complex sources such as: the multitude and invisibility of computer data, the technique of encryption, the difficulty

[17] Articles 20 and 21 of Council of Europe Convention on Cybercrime.

of identifying the perpetrator on the internet, the possibility that a computer can be attacked from a distance and the global nature of the internet that cannot be controlled by national borders. All these problems lead to lawmakers taking special measures for investigation of criminal activity on the internet. The Convention on Cybercrime provides legal rules for computer and network-specific investigation. Furthermore, other legal instruments[18] also provide special procedural rules addressing problems of international cooperation.[19]

According to Article 14 of the Convention on Cybercrime, the specific investigation procedures and procedural provisions of the Convention are applicable to all kinds of criminal activity on the internet. Article 14 states that:

> each party shall apply the powers and procedures of Section 2 of the Convention on Cyber-crime to (a) the criminal offences established in accordance with Article 2 through 1 of this convention; (b) other criminal offences committed by means of a computer system: and (c) the collection of evidence in electronic form of a criminal offence.

Put another way, the Convention on Cybercrime does not only apply to computer-specific offences determined in the Convention, but also to all "other criminal offences committed by means of a computer system" and to the "collection of evidence in electronic form of a criminal offence".[20]

These two subsections guarantee the special methods of investigation in the CoE Convention on Cybercrime that have the potential to address all kinds of criminal activities on the internet. Likewise, there are no gaps in applying the existing computer-specific procedural provisions of the Convention on Cybercrime to investigate cyberterrorism and other forms of terrorist use of the internet.[21]

---

[18] Such as Mutual Legal Assistance, Letter of Rogatory.

[19] Centre of Excellence Defence against Terrorism, *Cyber Terrorism — The Use of the Internet for Terrorist Purposes*, Council of Europe Publishing, France, 2007, p. 81.

[20] Council of Europe Convention on Cybercrime, Article 14.

[21] Centre of Excellence Defence against Terrorism, *Cyber Terrorism — the Use of the Internet for Terrorist Purposes*, p. 81.

This obliges parties to adopt legislative measures for computer-specific investigation. These cover the expedited preservation of stored computer data, the expedited preservation and partial disclosure of traffic data, search and seizure of stored computer data, interception of content data, real time collection of traffic data, safeguards for these measures, and jurisdiction rules. Thus, there are no limitations on using the existing computer procedural provisions of the Convention on Cybercrime for investigating cyberterrorism through the Convention. However, what is not certain is whether such instruments are adequate and up to date.[22]

Most of the investigation methods in the CoE Convention on Cybercrime are beneficial in investigating internet crime, both in general and terrorist cases. The Convention provides a "quick freeze procedure" which is required for the implementation of common trace back. This can be used if the perpetrator implements DDoS attacks via a number of third party computers as intermediaries in order to shield the information from other computers. The process of identifying the attacker depends on the analysis of the traffic data. However, such data is located in many computer systems in numerous countries and is not stored by the ISP; therefore, the need of a "quick freeze procedure" is vital because otherwise, the data would be erased. Moreover, the Convention provides specialised instruments in search and seizure of connected computer systems, or production orders to submit specified computer data because such data is difficult to investigate due to the encryption of data or IT applications' technical complexity which may be unfamiliar to the investigators.[23]

Related conventions on terrorism include specific and general rules on cyberterrorism investigation. The CoE Convention on the Prevention of Terrorism, the UN Convention against Transnational Organised Crime and UN Security Council Resolution 1373 have general procedural provisions with respect to terrorist acts and

---

[22] Centre of Excellence Defence against Terrorism, *Cyber Terrorism — The Use of the Internet for Terrorist Purposes*, p. 81.

[23] Centre of Excellence Defence against Terrorism, *Cyber Terrorism — The Use of the Internet for Terrorist Purposes*, p. 82.

offences. According to national procedural law, these conventions obligate their member states to undertake "the establishment of certain conditions and safeguards, the protection of victims of terrorism, the establishment of jurisdiction and the duty to investigate".[24] They also have specific provisions with respect to terrorism offences rather than merely criminalising the illegal content of terrorism activity. In addition, there are other regulations that are contained in the general instruments on mutual assistance and extradition, such as the European Convention on Mutual Assistance in Criminal Matters and its two additional Protocols, and the European Convention on Extradition and its two additional Protocols. The Convention on Cybercrime addresses the main problem of national procedural law with respect to cyberterrorism and terrorist use of the internet in all its forms.

Due to the fact that the general rules of investigation in these conventions have been drafted broadly, their provisions are applicable to cyberterrorism as well. According to Article 14 of the Convention on Cybercrime, the provisions are applicable to all criminal offences committed by means of computer systems. They apply to all types of terrorist use of the internet. Consequently, all existing international rules are applicable to cyberterrorism according to their national criminal procedures. Since these provisions apply whether or not offences are committed with the assistance of Information Technology (IT) they are ideal to be applied in an IT environment.

As a result, the above analysis shows that the procedural rules of the Convention on Cybercrime are applicable to all use of the internet for terrorist purposes, and even the problem of national procedural law is addressed by the Convention with respect to terrorist use of the internet. Therefore, in the next page, the Convention on Cybercrime, as the most significant tool in addressing the problem of cyberterrorism, is considered.

---

[24] Centre of Excellence Defence against Terrorism, *Cyber Terrorism — The Use of the Internet for Terrorist Purposes*, p. 83.

4.4.1 Investigation Process under the Cybercrime Convention

Investigation and prosecution when dealing with networks is inherently difficult, since obtaining the evidence and performing the investigation process in interconnected networks is almost impossible. Data and information can be deleted and altered as quickly as it is created.[25] Finding the identity of the cyber attack's perpetrator is the most difficult part of the investigation process, because perpetrators would attempt to shield their actual identity, and this can only be solved by cooperation among states.

To ascertain the perpetrator's identity, the first step is to examine the computer of the cyberterrorism victim to obtain the data which indicates the intrusion. Once found, the data, can be traced to identify the first router which sent the data, and then through that router it can be determined which network's server launched the cyber attack. Log files, data, and user IP addresses can be examined on the server's memory network. However, gathering this evidence in another country may cause problems, and thereby cause a conflict between national sovereignty and international collaboration.[26]

Article 14 of the Convention on Cybercrime states a general rule of investigation that is broadly drafted. The provision is applicable to cyberterrorism as well, since, according to Article 14, computer-specific investigation measures do not only apply to computer-specific offences defined in the Convention of Cybercrime, but also apply to other criminal offences committed by means of a "computer system"[27] and "collection of evidence in electronic form of a criminal offence"[28] as well.[29]

---

[25] P. Brunst, 'Legal aspects of cyber terrorism', In Centre of Excellence Defense against Terrorism, *Legal Aspect of Combating Terrorism*, IOS Press Publication, Turkey, 2008, p. 104.

[26] A. Reyes *et al.*, *Cybercrime Investigations Bridging The Gaps between Security Professionals, Law Enforcement, and Prosecutors*, Elsevier Publication, 2007, p. 87.

[27] Convention on Cybercrime (COC), e.g. Article 14.2(b) COC.

[28] Convention on Cybercrime (COC), e.g. Article 14.2(c) COC.

[29] P. Brunst, 'Legal Aspects of Cyber Terrorism'. In Centre of Excellence Defense against Terrorism, *Legal Aspect of Combating Terrorism*, IOS Press Publication, Turkey, 2008, p. 84.

It includes all methods of cyber terrorist activities and there are no gaps in applying the Convention rules on national criminal procedures to cyberterrorism. Cyber criminals cannot be prosecuted by law enforcement unless countries enact statutes which criminalise the offences related to terrorist activities in cyber space.[30] The Convention on Cybercrime in Articles 2–6, suppresses computer-related crime by establishing offences against the confidentiality, integrity, and availability of computer data and systems. The Convention on Cybercrime, by listing these offences, represents a consensus among member states in their domestic law.[31] However, due to the fact that the Convention on Cybercrime was drafted between 1997 and 2000, the technical methods and forensic investigation tools have changed. It appears that the procedural instruments of the Convention are not adequate to address suspected terrorism cases.[32] The necessity of an additional protocol should be explored for updating these procedural tools. The new protocol should consider the new risks posed by terrorism, new technical and forensic investigations and development.

Regarding the fast-paced technological cyber crime environment, these kinds of evaluation and upgrading are essential processes when high-risk threats such as cyberterrorism occur. Additionally, the serious threats coming from terrorist acts are not covered in an adequate manner by the Convention. In this regard, the protocol should focus more on preventative measures to act against "the dissemination of illegal content" in cyber space. In the same way, preventative measures play a role in respect to civil liberty and data protection as well.[33]

These developments can be followed by adding specific control systems on the internet to not prohibit the free exchange of

[30] Marc D. Goodman *et al.*, 'The emerging consensus on criminal conduct in cyberspace', *ULCA Journal of Law and Technology* (2002) 5(3), p. 2.

[31] Explanatory Report of Council of Europe on Convention of Cybercrime, Preamble Convention on Cybercrime, Budapest, 2001.

[32] Convention on Cybercrime (COC). Article 14.2(c) COC.

[33] Centre of Excellence Defense against Terrorism, Legal and policy evaluation: International coordination of prosecution and prevention of cyber terrorism, in *Responses to Cyber Terrorism*, NATO Advanced Research Workshop on Responses to Cyber Terrorism, Council of Europe, 4–5 October 2007, p. 91.

information without a reason. However, to gain complete fulfilment of the Convention's aims, it is necessary for the member states to enact domestic statutes to provide proper sanctions against illegal access and data and information interception. That is, the Convention on Cybercrime requires decisive sanctions in order to fulfil the intent of Convention — which is for the benefit of the member states — but the Convention authorises its member states to delineate such sanctions in their legislation, within the guidelines of the Convention on Cybercrime.[34] The cases of large-scale attacks in Estonia and Georgia taught the world that identifying the origins of an attack will require the cooperation of other states. This will not be achieved unless states are obliged through international law or duty to cooperate with other states in the investigation process.[35]

As one of the problems relating to prosecuting cyberterrorism cases is the jurisdiction issues, which is peculiar to instances of computer crime and cyberterrorism as one part of cyber crime, the principle of jurisdiction in the Convention on Cybercrime must be considered. Article 23 of the Convention establishes jurisdiction for the offences enumerated in Articles 2–11 of the Convention. The prescribed jurisdiction based on the Convention is a territorial and nationality jurisdiction. The Convention obliges member states to implement these criteria in their national legislation.[36] The most common principle of territorial jurisdiction is that a sovereign state has the authority to assert jurisdiction over criminal acts that have been committed in its territory; however, according to the ubiquity doctrine, in many cases the extraterritorial effect may apply on the territoriality principle.[37] Nevertheless, territorial jurisdiction cannot be implemented in cyberterrorism cases because cyberterrorism operates without respect to any

---

[34] Centre of Excellence Defense against Terrorism, Legal and policy evaluation: International coordination of prosecution and prevention of cyber terrorism, in Centre of Excellence Defense against Terrorism. *Responses to Cyber Terrorism*, IOS Press Publication, Turkey, 2007, p. 91.

[35] C. E. Lentz, 'A state duty to prevent and respond to cyber terrorist acts', *Chicago Journal of International Law* (2010) 10(2), p. 417.

[36] Convention on Cybercrime (COC), Article 23 COC.

[37] H. W. Kaspersen, Discussion Paper on Cybercrime, and Internet Jurisdiction, Council of Europe Human Rights (2009), p. 9.

border and ignores the parameters that have been erected by a state law. Furthermore, cyberterrorism occurs in cyber space and involves attacks not only on specific data, but also targets whole computer systems, such as infrastructure and power grids.[38]

More than this, tracing the identity and location of cyber terrorists is nearly impossible, due to the fact that cyberterrorism uses new technological tools such as spoofing by using an anonymous IP address and DoS attacks to conceal the real identities and to pretend that the attack came from somewhere else. Cyber terrorists commit their crimes by using computers in multiple states.[39] Smart cyber terrorists route their attacks through third-party networks. They exploit countries whose governments have poor diplomatic relations as well as no law enforcement. They obscure the true origin of an attack by providing administrator authority over a series of compromised computers to reach their target without the awareness of the current computer user.[40] However, the Convention on Cybercrime establishes jurisdiction over criminal offences in order to condemn and prosecute cyber crime to have a deterrent effect. Applying territorial and nationality jurisdiction is impractical in combating cyberterrorism since those jurisdictions cannot perform this primary function of deterrence on their own.

The procedural law section in the CoE's Convention on Cybercrime is largely based on the CoE Recommendation of 1995 regarding the problems of criminal law connected with IT. Effective investigation and prosecution at an international level requires the establishment of precise and general procedural regulation in collecting, preserving, and presenting evidence in electronic form. Furthermore, mandatory data retention for ISPs for a period is required for law enforcement investigation. This strategy is legalised in some EU countries such as the UK, Sweden, France, and Ireland.

---

[38] K. A. Gable, 'Cyber-apocalypse now: Securing the internet against cyber terrorism and using universal jurisdiction as a deterrent', *Vanderbilt Transnational Law* (2010) 43, p. 73.

[39] K. A. Gable, 'Cyber-apocalypse now: Securing the internet against cyber terrorism and using universal jurisdiction as a deterrent', p. 74.

[40] K. Geers, 'The challenge of cyber-attack deterrence', *Computer Law and Security Review* (2010) 26, p. 301.

States must adopt measures for law enforcement bodies to order or obtain expedited preservation of stored computer data and expedited preservation and partial disclosure of traffic data.[41] The existing data should not be altered or deleted by ordering a service provider, unless disclosure is obtained. Consequently, the service provider must preserve the integrity of the computer data for a period of up to a maximum of 90 days. This could be done by obliging the data-holder to preserve and expeditiously disclose enough traffic data to enable the identification of the service providers and the path through which the communication was transmitted. If not, traffic data may be lost and law enforcement bodies would not be able to track a communication back to its source, especially where several service providers are involved.

States must have the authority to access a person's computer data or computer data storage medium in their territory. Such measures should be adopted by all states. Alternatively, the authorities must have the ability to order the service provider to submit stored subscriber information (any information on subscribers in the form of computer data).

Another measure which must be adopted by states is the search and seizure of stored computer data. This measure is the same as the legal authority of search and seizure in tangible property; only the environment is different. The law enforcement body must have the authority to search the computer data and seize the computer system or the computer data storage medium that maintains a copy of the computer data and the integrity of the stored computer data by using computer equipment or electronic communication systems and simultaneously, make it inaccessible or remove it.

4.5 THE INVESTIGATION PROCESS IN CYBERTERRORISM: AN ANALYSIS

The fact is that cyberterrorism investigation is surrounded by complexity and this complexity hinders the investigators because the internet provides the opportunity for cyber criminals to react quickly.

---

[41] H. W. Kaspersen, *Discussion Paper on Cyber-crime and Internet Jurisdiction*, p. 15.

Cyber criminals utilise a distributed approach to conceal their activities globally and across several jurisdictions.[42] Law enforcement needs to react and keep pace with changes occurring through the different methods of cyber attacks. A key feature in conducting appropriate investigations and law enforcement is to manage information and knowledge among all parties involved. Like physical investigation, cyberterrorism investigation is information-rich and it is vital to gain a detailed insight into illicit data attacks rapidly.[43]

The research introduced by Hunton highlights that it is essential for investigators to understand cyber attacks, and the data objectives, exposition tactics, methods, and technical implementation of an attack. The illicit intent of the cyber criminal is defined by the data objectives (data collection, data supply and distribution, data use).[44] This illicit intent forms a major part in investigating cyberterrorism, because one of the key distinctive differences between cyberterrorism and other cyber crimes is the illicit intent of the cyber criminal which must be political, religious, or similar. By identifying the illicit intent of the cyber criminal, their technical activities can emerge and be prevented. Furthermore, it is useful to consider the attack based on the cyber criminal's exploitation tactics: attacking targeted devices with the intention of using technology for illicit purposes, using network technology as a mechanism, and deception attacks and data. This will bring the investigator to the technical methods and intent of a cyber attack and allow them to distinguish the technology methods utilised in the cyberterrorism attack. This is because cyberterrorism involves a combination of different attack methods and as a result, the technology resources used for large-scale data will be very different.[45]

---

[42] P. Hunton, 'Data attack of the cybercriminal: Investigating the digital currency of cybercrime', *Computer Law and Security Review* (2012) 28, p. 204.

[43] P. Hunton, 'Data attack of the cybercriminal: Investigating the digital currency of cyber-crime', p. 203.

[44] C. Police, 'A rigorous approach to formalizing the technical investigation stages of cyber-crime and criminality within a UK law enforcement environment', *Digital Investigation* (2011) 7, p. 107.

[45] C. Police, 'A rigorous approach to formalizing the technical investigation stages of cybercrime and criminality within a UK law enforcement environment', p. 108.

It appears that data distribution is applied by cyber terrorists when deploying large-scale cyber attacks. The data collection for an attack can be done in various ways. In cyberterrorism, this data collection is conducted through large-scale data attacks, such as email spam and phishing websites. Cyber terrorists mostly utilise data distribution that applies on a large scale, such as "botnets". A survey showed that in the first 6 months of 2010 in the US more than 2.2 million computers were infected through distributed attack methods such as spam, emails, or data flood attacks.[46]

In common law countries like UK and Malaysia, prosecutors must decide what cases they take on, while in other legal systems, prosecutors are required to prosecute when sufficient evidence is available. These decisions are reached by the prosecution policies in some modern jurisdictions based upon the seriousness of the offence and the sufficiency of the evidence, but in others, the law dictates it. Prosecutors in common law jurisdictions must present the evidence in court; in other jurisdictions they just assist the judge in fact finding.[47] The first step after providing the evidence in court is imposing the appropriate sentence to a convicted offender. In order to impose sentence, the principle of *mullum crimen sine lege* must be observed. Under this principle, the behaviour cannot be prosecuted without law that prohibits the behaviour. This can be seen in the case of the "I Love You Virus" that was disseminated in many jurisdictions and originated from the Philippines. Significantly, the Philippines government could not prosecute the offender, since they did not have any law at that time which prohibited the release of malicious code. The same thing happened in the Estonian case.[48] In such cases, if the criminal law of that country has sufficient flexibility, it can prosecute conventional crimes committed via digital technology.[49]

---

[46] Microsoft, *Battling Botnets for Control of Computers Microsoft Security Intelligence Report*, 2010, pp. 12–16.

[47] S. Brenner *et al.*, 'Transnational evidence gathering and local prosecution of cybercrime', *John Marshall Journal of Computer and Information Law* (2002) 20, p. 347.

[48] S. Brenner *et al.*, 'Transnational evidence gathering and local prosecution of cybercrime', p. 347.

[49] P. Grabosky, 'Requirements of prosecution services to deal with cybercrime', *Crime Law Social Change* (2007) 47, pp. 208–209.

Technology advances faster than law; therefore, the legislation must be broad enough in order to be capable to embrace future technological development, while not being so general as to be vague. The ideal way is to use technology-neutral language. In the case of *R v. Fellows and Arnold* [50] the defendant was charged with possessing indecent photographs under the Protection of Children Act 1978. The defence argued that a photograph in digital form did not constitute a photograph. Nevertheless, the court held that they were photographs but in a different form.[51]

In criminal procedure law, one of the significant stages is searching and seizing computers. The search power must be in accordance with the rule of law. Some countries do not admit the evidence obtained from searches that exceed the scope of a warrant. Therefore, the prosecutor must seek, draft, and execute it in a proper manner.[52] The search of the prosecutor must be within the scope of the warrant.[53] Various countries have different approaches. The digital evidence resulting from the search warrant enables the prosecution and conviction of the criminal.[54]

4.5.1 Gathering Evidence and Prosecuting through Formal and Informal Forensic Investigation

Although new types of crimes are emerging in the cyber world, law enforcement still relies on traditional methods of investigation and prosecution of such new crimes. Such new types of crimes require new laws and digital forensic investigation to enable the digital evidence to be presented in a court of law. However, the digital evidence is presented in the court of law and if the procedures were conducted

---

[50] R v. Arnold [1997] 1 Cr App R, p. 244.

[51] R v. Fellows and Arnold [1997] 1 Cr App R, p. 244.

[52] E. Drozdova, 'The Transnational Dimension of Cyber-crime and Terrorism', Hoover Institute Press, Stanford, p. 200. Available at: http://www.hoover.org/publications/books/cybercrime.html (23 Apr 2012).

[53] S. Brenner *et al.*, 'Transnational evidence gathering and local prosecution of cybercrime', *John Marshal Journal of Computer and Information Law* (2002) 20, p. 347.

[54] A. Khan *et al.*, 'Digital forensics and crime investigation: Legal issues in prosecution at national level', *IEEE Computer Society*, p. 244.

based on the step-by-step procedure in cyber and electronic laws, it will be admissible in the court of law.

Evidence collected from digital devices has a very important role in prosecution. The reporting mechanism of transporting evidence is very important to report the crimes to a law enforcement agency. Then, the law enforcement agency must respond to the complaint immediately to trace the incident back to the perpetrator. Put simply, as soon as a complaint is lodged at the National Response Center for Cybercrime, the helpdesk officer will ascertain the nature of the complaint and forward it by using an online system to relevant field units for proper action. The relevant field levels will process the complaint with the help of forensic experts, legal advisors, and take appropriate legal action. The forensic expert usually attempts to maintain the originality and integrity of the collected digital evidence by using the most suitable method of protecting the evidence and identifying methods for recovery, preservation, and presentation of digital evidence in the court of law. On the other hand, the legal advisor will attempt to ensure that the procedures are in compliance with existing legal requirements and provide legal counsel for cyber crime investigation.

Finally, each State must have a complete national law to investigate and prosecute cyberterrorism as a part of cyber crime. As far as the conventional law does not cover cyber crimes, the new law must be broad enough to cover the latest cyber crimes, such as cyberterrorism. This law also must include cross-border jurisdiction as well as extradition agreements or MLATs in order to include new types of cyber crimes, such as cyberterrorism.

Generally, cyber crime investigation is conducted through two alternative forms: informal cooperation and formal mechanisms.[55] International cooperation and mutual legal assistance play an important role in digital investigation and in the prosecution process, especially in cyberterrorism cases which occur in different countries. This is because the footprints of such crimes are fragile and in transition;

[55] S. W. Brenner *et al.*, 'Introduction — cybercrime: A note on international issues', *Information System Frontiers* (2004) 6(2), p. 111.

therefore, traditional law enforcement and investigation are not adequate. Due to the multiple countries involved and the formal processes involved in countering loss of evidence in order to commence the investigation, the investigator is required to obtain the assistance of the authorities of the country where the crime originated, or the countries through which the activity transited. Consequently, the investigator must determine which devices are available for search and seizure of evidence which is located in other countries. Formal mutual assistance includes bilateral mutual assistance treaties, while multilateral mutual assistance is required obtain the permission of prosecution and investigation processes, or letters rogatory in the absence of a treaty or executive agreement. The procedure of MLATs is faster than the old process of letters rogatory.[56,57]

Following the identification of the hacker, the next and most important part is to bring the hacker to trial. The US, as one of the most-targeted countries, provides two methods for bringing the citizen-hacker to the US which are the extradition treaty and extralegal seizure. Usually, the US does not enter into treaties with countries with repressive regimes, and therefore there will be no legal grounds for the US to enforce extradition treaties with countries with which it does not have reciprocal treaties. For example, the US does not have any reciprocal treaty with China and therefore in circumstances where an incident happens in the US, cannot bring the perpetrator to trial in this way. In *Factor v. Laubenheimer*,[58] the Supreme Court held that the legal right for a country to demand extradition of another country's citizen is only applicable by treaty. On the other hand, it is doubtful that an extradition would occur even if a treaty existed between two countries. An extradition treaty usually includes a political offence exception.

Generally, attacks originating from American citizen-hackers, either on a vital target or non-vital target, would be controlled by the Computer Fraud and Abuse Act. Thus, whenever attacks emanate

---

[56] Letter of Rogatory explained in Chapter III.

[57] P. Grabosky, 'Requirements of prosecution services to deal with cyber-crime', *Crime Law Social Change* (2007) 47, pp. 214–215.

[58] Factor v. Laubenheimer [1933] United States Supreme Court, p. 293.

from a private citizen or non-state actor, any response is subject to law enforcement issues, i.e. even when the non-state actor and state actor causes identical damage to a state's infrastructure, the actions conducted by the non-state actor remain bound by law enforcement issues.[59]

The process of MLATs in the US is as follows. The investigator should first determine whether any MLAT exists between its country and the country where the evidence is located. The US has MLATs in effect with many countries. "If an MLAT encompassing the evidence in question is in effect, the investigator prepares a request for assistance pursuant to the treaty; such a request contains essentially the same information as that used for a letter rogatory, except for the promise of reciprocity".[60] In the absence of an effective MLAT, the investigator will have to resort to the letter rogatory process. The judiciary of a foreign country, on the request of a US' judge, issues the letter rogatory. It is the duty of the investigator to fulfil the letter rogatory and submit to the authority of the country which requests the assistance. The letter rogatory must have the following requirements:

> (a) the names and affiliations of the officers conducting the investigation; (b) enough facts about the case for a foreign judge to understand that a crime has been committed and see the relevance of the evidence sought; (c) the nature of the assistance required from the foreign authorities; (d) the text of the statutes alleged to have been violated, and (e) a promise of reciprocal assistance.[61]

However, obtaining assistance through a letter rogatory may take a year or more to accord with the wishes of the Department of Justice, because it is transmitted through diplomatic channels which make it a time-consuming method.[62]

---

[59] W. G. Sharp, *Cyberspace and the Use of Force*, Aegis Research Corporation, United States of America, 1999, pp. 154–156.
[60] S. W. Brenner, 'Introduction — cybercrime: A note on international issues', p. 112.
[61] U.S. Department of Justice, U.S. Attorney's Manual, Section 275.
[62] U.S. Department of Justice, U.S. Attorney's Manual, Section 275.

Police officers traditionally use the informal method. Due to the rise of cyberterrorism and lack of knowledge of investigators with law enforcement bodies in other countries, they need assistance. Therefore, the informal method is more effective if supplemented by the efforts of international organisations such as the COE and the G-8 Network of 24-Hour Points of Contact for High-Tech Cases in order to be effective. Thus, informal assistance that is based upon good relationships is the proper approach, since it removes the administrative procedure of approval.[63] In fact, investigators can receive information about how to obtain evidence from abroad from international sources of guidance such as Interpol and the G8. US investigators can also obtain such information from the US Department of Justice and the National Institute of Justice International Center.[64] However, the informal method is not as reliable as expected.

A good example of such kind of evidence gathering is the Love Bug Virus case in which the problem originated from the inconsistencies that existed among nation's laws. As soon as the virus spread throughout the world, the US traced the virus to the Philippines. The FBI cooperated with local law enforcement officers to investigate the house of Onel du Guzman, who was the primary suspect. Informal cooperation worked at its best, but the problem was from the Philippines legal system because, the Philippines' criminal code did not criminalise the dissemination of viruses or hacking. Finally, the officers obtained a warrant and conducted the search.

Onel du Guzman was charged with fraud and theft, but the charges were dismissed, since according to the Philippines criminal code, hacking and dissemination of viruses were not criminalised. The US Department of State is a source that offers general information about obtaining evidence from abroad and letters rogatory, and lists all MLATs and executive orders which are currently in force.[65]

---

[63] P. Grabosky, 'Requirements of prosecution services to deal with cyber-crime', pp. 214–215.

[64] Law Enforcement & Counterintelligence Center. Available at: http://www.cert.mil/lecic/lecic.html (30 April 2012).

[65] S. W. Brenner, 'Introduction — cybercrime: A note on international issues', p. 114.

As a result, the conclusion of this discussion proposes that investigating and prosecuting of cyberterrorism and cyber crime offences requires cooperation between law enforcement officials in various countries. The only proper way to reach an internationally effective response to cyberterrorism is to harmonise domestic law to be consistent when dealing with cyber crime offences and in the procedures for evidence gathering for law enforcement in cyber crime investigation.[66]

The Invita case illustrates the approach of investigators to the gathering of evidence across national borders. This occurred following the intrusion of two Russian hackers who targeted ISPs, e-commerce sites, and online banks in the US. They used unauthorised access to the victims' computers to steal financial information and extort money from people by threatening to publish their personal information. The FBI identified them, used a ruse, enticed them to the US by creating a bogus company, and asked them to hack the Federal Bureau of Investigation (FBI) network as a part of their interview with the purpose of checking their computer skills.[67]

In doing so, they accessed their computer system in Russia. The FBI had installed a keystroke logger program on the laptops provided for the two, Gorshkov and Ivanov, recorded their usernames and passwords to access their Russian computers. Following that event, the FBI logged on to their computer in Russia, downloaded and copied the file contents of their computer without a warrant. Gorshkov and Ivanov argued that this evidence was the product of a search and seizure and violated the Fourth Amendment as well as Russian law.[68]

The court held that they did not meet the Fourth Amendment expectation of privacy because they used a foreign computer to access the files in Russia. The Fourth Amendment does not apply to search and seizure conducted outside the boundaries of the US, since the computer was located in Russia and it was a property of a non-resident. In addition, the principle of expectation of privacy could not apply to the defendants, because the computer belonged to a US Company

[66] S. W. Brenner, 'Introduction — cybercrime: A note on international issues', p. 114.
[67] Gorshkov, [2001] WL 1024026, p. 1.
[68] Gorshkov, [2001] WL 1024026, pp. 3–4.

and they knew that the system administrator had the ability to monitor their activities. Therefore, the court held that they did not have any reasonable cause for asserting expectation of privacy.

They believed that the FBI action was reasonable according to exigent circumstances because the agents feared that the evidence could be destroyed, or that it could be made unavailable by transfer to a different computer, or access prevented by changing the password. In this case, the law governing search and seizure was the law where the crime was committed, in this instance, the law of the US.

Another relevant case is the "Rome Labs" case at Griffiss Air Force Base in New York, where the network was hacked by a hacker calling himself the "DataStream Cowboy". He installed a password sniffer program in the entire Rome Labs network and used the Lab's system to attack other systems in multiple countries in South America and Europe, including Mexico and Hawaii. The Air Force's Office of Special Investigation (AFOSI) identified "DataStream Cowboy" as a 16-year-old citizen of the UK called Richard Pryce.

They contacted New Scotland Yard officers — with whom they had an existing relationship — to monitor Pryce's telephone lines. New Scotland Yard obtained a warrant to search Pryce's residence based on this information and executed the warrant and seized alleged evidence. Pryce was prosecuted and pleaded guilty to 12 counts of hacking. The Rome Labs investigation operated based on the premise that the law of the state where the property to be searched or seized was the governing law. The American agents ensured that the evidence in possession of the British was gathered in accordance with British law by contacting their counterparts in Britain and in cooperation with New Scotland Yard officers. This case is an excellent example of informal cooperation in investigation.

4.5.2 Evaluation of Evidence

Another issue that arises in this category is determining whether the conduct is illegal in the country whose assistance is being requested. The law of the requested party may differ from that of the requesting party and some countries only grant assistance when the conduct is also illegal under their domestic laws. Alternatively, they may treat the conduct as a normal offence, while the country that seeks the assistance treats it as a capital offence. Investigators and prosecutors face various problems like these.

4.6 CURRENT PROSECUTION PROCESS IN CYBERTERRORISM CASES

Generally, obtaining evidence from another country is troublesome in cyber crime cases. The prosecutor must learn what evidence is admissible in other jurisdictions as well as how to obtain this evidence by formal and informal methods.

4.6.1 Transnational Evidence and the Prosecutor: Current Challenges

The first problem in transnational cyber evidence is the lack of resources to acquire such data. The evidence acquired must be logically and legally relevant. Aside from this problem, a local prosecutor who is involved in the case must go a long way within the international context. For e.g. if the local prosecutor needs to obtain an IP in Italy, he or she must make a request to the OIA and Department of Justice and they will engage the procedures of the MLAT between the US and Italy to ascertain what may be done. This would take several months and may not lead to punishment of the criminal because the charges may not have been filed, or the evidence may not be sufficient or be dismissed because of the delay. Also, the government has to spend a huge amount of money to obtain the data. Due to these facts, it is said that informal assistance is better, since it saves time, money, and the chances of sentencing the criminal are increased.

4.6.2 Search Warrant: An Important Tool for the Prosecutor

In criminal procedure law, one of the significant stages is searching and seizing computers. The search power must be in accordance with the rule of law. Some countries do not admit the evidence obtained from searches that exceed the scope of a warrant. Therefore, the prosecutor must seek, draft, and execute it in a proper manner.[69] The search warrant is applied for in two ways in different countries. In some countries the law enforcement officers must obtain judicial authority before searching and seizing, while in others this authority is vested in prosecutors or police. To obtain a search warrant, the requesting party or investigator must provide sufficient cause that a crime has been committed and that the electronic device located at the place was involved in committing the crime. Then the investigator must obtain a warrant from the court for search, seizure, or arrest of a suspect. In addition, he or she must determine with considerable specificity what evidence must be sought. Some countries provide a general search warrant while others, such as the US, require specific details about what is to be searched and sought as evidence. Whenever the prosecutor begins the search, it must be within the scope of the warrant.[70] The warrant details "the intrusiveness of proposed search and the location of the computer". Various countries have different approaches. The digital evidence resulting from the search warrant enables the prosecution and conviction of the criminal.[71]

4.6.3 Search Warrants in Cyberterrorism Cases: The US Experience

According to the US' Fourth Amendment of the Constitution, in order to obtain a criminal warrant, a federal agent must prove that under the circumstances known to him or her there is a reasonable belief that a person has committed, is committing, or is about to commit a crime. According to the Electronic Communications Privacy Act (ECPA) 1986, it is illegal to use any form of electronic communication

[69] E. Drozdova, 'The transnational dimension of cybercrime and terrorism', Hoover Institute Press, Stanford, p. 200. Available at: http://www.hoover.org/publications/books/cybercrime.html (23 Apr 2012).
[70] S. Brenner *et al.*, 'Transnational evidence gathering and local prosecution of cybercrime', p. 347.
[71] A. Khan *et al.*, Digital forensics and crime investigation: Legal issues in prosecution at national level, p. 244.

to convict or accuse someone of a crime without obtaining a prior search warrant. It is also illegal to listen in on communications to obtain a search warrant, or to interrupt transmissions. For instance, according to the ECPA, police or the government cannot use a tapped phone unless they have obtained a search warrant. Therefore, if a phone has been tapped without authorisation, the suspect can voice his/her disagreement, and consequently this evidence cannot be used. Moreover, a private taped conversation cannot be submitted as evidence if the citizen's consent has not been obtained.[72]

The first thing in Title I of the ECPA is the definition of electronic communication. Any intentional and attempted interception of electronic communication and the use of illegally-obtained electronic communication are prohibited under the ECPA. In Title II, the Act also prevents hackers from obtaining, altering, or destroying certain stored electronic communications. It can also be used on anyone who "intentionally accesses without authorization a facility through which an electronic communication service is provided; or intentionally exceeds an authorization to access that facility; and thereby obtains, alters, or prevents authorised access to a wire or electronic communication while it is in electronic storage in such system shall be punished".[73] For the prosecution of computer-related electronic communication violations, the ECPA and Federal Wiretap Act work together and the effect of this conjunction allowing prosecutors to adapt with changes in technology. The USA Patriot Act (2001) changed the wiretap statutes in two ways: "by adding felony violations of the computer hacking statute to the list of predicate offenses for the interception of communications; and by changing the way in which the Federal Wiretap Act and the ECPA apply to stored voice communications, allowing federal agents to obtain protected communications under the less demanding procedures of the ECPA

[72] H. M. Jarrett *et al.*, *Searching and Seizing Computers and Obtaining Electronic Evidence in Criminal Investigations*, 1st Edn, Office of Legal Education Executive Office for United States Attorneys Publication, United States of America, 2009, pp. 15–17.

[73] 18 U.S.C. § 2701(a)(2000).

rather that the more demanding wiretap order required by Section 2516".

ECPA prevents governments from circumventing their positions in cyberterrorism cases by handing over private conversations of people without a warrant. Otherwise, they may hand over these private conversations in case they doubt the commission of cyberterrorism over specific conversations. The amendments included in the Patriot Act of 2001 mostly covered existing federal laws. These changes reference the Computer Fraud and Abuse Act[74] in both procedural and substantive changes that may significantly influence future prosecution. These changes make investigation of computer crime easier for law enforcement officers.[75]

As discussed above, the Patriot Act (2001) contains two procedural modifications that relate primarily to computer activity. First, it specifically adds felony acts related to the Computer Fraud and Abuse Act to the list of grounds that authorise the right to intercept wire, oral, and electronic communications. A second procedural change that relates directly to computer crimes is the section that authorises the interception of computer trespasser communications. It grants federal officials greater powers to trace and intercept terrorists' communications for dual purposes: law enforcement and foreign intelligence. The authority to monitor ranges from email traffic, to sharing grand jury information with intelligence and immigration officers, to confiscating property, to imposing new book-keeping requirements on financial institutions.[76] This Act is the main national counter-terrorism tool and expands the government's ability to monitor Americans in the name of national security.

Furthermore, it expands government investigatory authority over online information. It strengthens capabilities in several areas, including surveillance, search warrants, detention, restricted access, money

---

[74] 18 U.S.C. § 1030.

[75] E. S. Podgor, 'Computer Crime and the USA Patriot Act (2001)' (2002) 17(2), *Criminal Justice Magazine*. Available at: http://american bar.org/publications/criminal justice magazine home/crimejust cjmag 17 2 crimes.html (24 May 2012).

[76] C. Doyle, The USA Patriot Act (2001): A Legal Analysis, RL 31377, CRS Report of Congress American Law Division, p. 5.

laundering, information sharing, and criminal penalties.[77] It was passed in response to the terrorist attacks of 11 September, 2001,[78] and a part of this Act is an amendment to federal surveillance law, which governs the capture and tracking of suspected terrorist communications within the US. Federal law gives the authority to identify and seize criminal communications from private telephones, face to face and computer communications. In other words, it provides a three-tiered system. This can be seen in the Fourth Amendment of the Supreme Court which bans unreasonable seizure.

After 11 September 2001 the Inspector General of the Department of Justice appointed an official to monitor, review, and report back to Congress on all allegations of civil rights abuses against the Department. According to Title III, the court order described the legitimate duration and scope of surveillance and seized conversation. Following the expiration of the court order, the court notified the parties involved in the seized conversations.

The government may use trap and trace devices and pen registers in order to identify the source and endpoint of calls which were made to and from specific telephones, and the orders to allow this can be obtained by government certification.[79] However, the Amendment does protect private conversations.[80] It does allow the cloaking of even highly personal information for which there is no individual justifiable expectation of privacy such as an individual's telephone company records,[81] or bank records.[82] Congress responded to the case of *Berger and Katz* by passing Title III of the Omnibus Crime Control and Safe Streets Act of 1968.[83] Title III, as amended, prohibits electronic

---

[77] C. Doyle, the USA Patriot Act (2001): A Legal Analysis, p. 5.

[78] Congressional Research Service, *The USA Patriot Act: A Legal Analysis,* Cyber telecom Federal Internet Law & Policy an Educational Project, 2002. Available at: http://www.cybertelecom.org/security/patriot.htm (21 May 2011).

[79] C. Doyle, the USA Patriot Act (2001): A Legal Analysis, p. 4.

[80] *Berger v. New York*, 388 U.S. 41 (1967); *Katz v. United States*, 389 U.S. 347 (1967).

[81] *Smith v. Maryland*, 442 U.S. 735 (1979).

[82] *United States v. Miller*, 425 U.S. 435 (1976).

[83] 18 U.S.C. 2510–2522 (Title III).

eavesdropping on telephone conversations, face-to-face conversations, or computer and other forms of electronic communications,[84] while, at the same time, providing authorities with a strictly-defined process for electronic surveillance, which is only to be used as a last resort in serious criminal cases. After seeking the approval of senior officials of the Justice Department, law enforcement officers may seek a court order authorising them to secretly monitor conversations relating to any of a statutory list of offences (the predicate offences).[85]

As a result, this Act also authorises nationwide execution of court orders for pen registers, trap and trace devices, and access to stored email or communication records. The Patriot Act (2001) permits nationwide or even worldwide search warrants in terrorist cases. The Act allows court orders authorising trap and trace devices and pen registers to be used to capture sources and addresses information in computer, email, and telephone conversations.[86] Three sections of the Patriot Act (2001) that are in force will:

(1) Authorise court-approved roving wiretaps that permit surveillance on multiple phones.
(2) Allow court-approved seizure of records and property in anti-terrorism operations.
(3) Permit surveillance against a so-called lone wolf, a non-US citizen engaged in terrorism who may not be part of a recognised terrorist group.[87]

The law permits criminal investigators to retrieve the content of electronic communications in storage, such as email, by search warrant. The duration can reach up to 180 days without notifying the subscriber, if the communication has been in remote storage.[88] It also

[84] 18 U.S.C. 2511.
[85] 18 U.S.C. 2516.
[86] 18 United States Code 2703.
[87] 'anon' 'Obama signs 1-year extension of Patriot Act', *Fox News*, 27 February 2010. Available at: http://www.foxnews.com/politics/2010/02/27/obama-signs-year-extension-patriot-act/ (20 Sept 2011).
[88] 18. United States Code 2703(a)(b).

covers seizing records describing telephone and other communication transactions.

Officers may resort to a court order to access electronic communication in situations in which such information is relevant to a criminal investigation for more than 180 days, with no need to notify the subscriber.[89] Sometimes, there is no notification at all, since any notification would jeopardise or may lead to undue delay of the trial. The accessibility of this information operates on the principle of 'necessity'; it prohibits the seizure of any tangible property or electronic communication except when the court finds it a reasonable necessity.[90]

Moreover, according to Section 220 of the Patriot Act (2001), the jurisdictional restrictions on access to the content of stored email pursuant to a court order were eliminated.[91] To streamline the investigation process, the Act allows credit card and bank account number information to be subpoenaed by law enforcement officials from customer records kept by the customer service provider.[92] Finally, the government may have access to cable company customer records only on the basis that they can show that the records will form evidence that the customer is or has engaged in criminal activity.[93]

Section 217 permits law enforcement officials to intercept the communications of an intruder within a protected computer system without the necessity of a warrant or court order. As a matter of fact, this provision defines all the related issues, such as a 'protected computer', 'Department of the United States', 'states', 'financial institution', 'exceeds authorized access', 'damage', etc. According to this Act, Congress eliminated all mandatory minimum guidelines on sentencing which existed before. In the previous law, any violation under the Computer Fraud and Abuse Act entailed a mandatory sentence of at least 6 months' imprisonment.

---

[89] 18. United States Code 2703(b)(c).

[90] 18 U.S.C. 2703(c).

[91] 18 United States Code 2703.

[92] 18 United States Code 2703 (c)(1) (C).

[93] 47 United States Code 511 (h).

The Patriot Act (2001) increases the penalties for acts of terrorism and the crimes which terrorists might commit. It establishes an alternative maximum penalty for acts of terrorism, raises the penalties for conspiracy to commit certain terrorist offences, envisions sentencing some terrorists to life-long parole, and increases the penalties for counterfeiting, cyber crime, and charity fraud.[94] According to suggestions of the Justice Department, anyone who is convicted of an offence designated a terrorist crime must be given the penalty of imprisonment, up to life imprisonment. Put simply, it increases the maximum penalties of various crimes of terrorism. Therefore, it increased the maximum terms of imprisonment for several crimes:

(1) For life-threatening arson or arson of a dwelling committed within a federal enclave, from 20 years to any term of years or life[95];
(2) For causing more than US$100,000 in damage to, or significantly impairing the operation of an energy facility, from 10 to 20 years (or any term of years or life, if death results)[96];
(3) For providing material support to a terrorist or a terrorist organisation, from 10 to 15 years (or any term of years or life, if death results)[97];
(4) For destruction of national defence materials, from 10 to 20 years (or any term of years or life, if death results)[98];
(5) For sabotage of a nuclear facility, from 10 to 20 years (or any term of years or life, if death results)[99];
(6) For carrying a weapon or explosive abroad an aircraft with US special aircraft jurisdiction, from 15 to 20 years (or any term of years or life, if death results)[100]; and
(7) For sabotage of interstate gas pipeline facilities, from 15 to 20 years (or any term of years or life, if death results).[101]

---

[94] C. Doyle, The USA Patriot Act (2001): A Legal Analysis, p. 58.
[95] 18 U.S.C. 81.
[96] 18 U.S.C. 1366.
[97] 18 U.S.C. 2339A, 2339B.
[98] 18 U.S.C. 2155.
[99] 42 U.S.C. 2284.
[100] 49 U.S.C. 46505.
[101] 49 U.S.C. 60123.

Finally, the Patriot Act (2001) increases the penalties for acts of terrorism and for crimes which terrorists might commit. It establishes maximum penalties for acts of terrorism and increases the penalties for conspiracy to commit certain terrorist offences, counterfeiting, cyber crime, and charity fraud.

A. Search warrant for obtaining foreign intelligence information in the US

Title II of the Patriot Act covers all aspects of the surveillance of terrorists suspected to be engaged in clandestine activity. This Title includes the amendment of the Foreign Intelligence Surveillance Act (FISA) of 1978 amendments act of 2008 and Electronic Communication Privacy Act (EPIC). It allows the government to gather "foreign intelligence information" from both US and non-US citizens, and changed FISA to make gaining foreign intelligence information the significant purpose of FISA-based surveillance, where previously it had been the primary purpose. According to FISA, the sole purpose of a FISA warrant was that foreign intelligence affected national security and it was confined to agents of a "foreign power". In cyberterrorism cases, terrorists attempt to affect national security through their destructive attacks. FISA warrants are thus beneficial for prosecuting cyberterrorism cases.

To apply for a surveillance order under FISA requires a certification of the fact that "the purpose for the surveillance is to obtain foreign intelligence information".[102] According to the language of Section 1802 of US Code, as well as the requirement of Section 1804 of the US Code, the application must "contain a certification by a designated official of the executive branch that the purpose of the surveillance is to acquire foreign intelligence information, and the certification must set forth the basis for the certifying officials' belief that the information sought is the type of foreign intelligence information described".[103]

---

[102] 50 U.S.C. 1804(a)(7)(B).

[103] C. Doyle, The USA patriot act (2001): A legal analysis, RL 31377, p. 9.

In *United States v. Pelton*,[104] Pelton's claim was rejected because the FISA surveillance was conducted primarily for the purpose of his criminal prosecution, and not primarily for obtaining foreign intelligence information, and the previous District Court decided that the primary purpose of the surveillance was to gather foreign intelligence information. Although the gathering of evidence by FISA may be used subsequently in the criminal prosecution, the primary purpose of the surveillance cannot be the investigation of criminal activity. FISA was required to be used where the only purpose of the investigation was to gather foreign intelligence.

However, under the new amendment in the Patriot Act (2001) concerning foreign intelligence investigations, foreign intelligence need only be "a significant purpose of the surveillance". It expands the definition from agents of a "foreign power" to persons associated with "a group engaged in international terrorism or activities in preparation".[105] Therefore, it can include American citizens who are not associated with a foreign power as the targets of a FISA warrant.

Senator Dianne Feinstein believes that such "changes were necessary to make it easier to collect foreign intelligence information under FISA. Under current law authorities can proceed with surveillance under FISA only if the primary purpose of the investigation is to collect foreign intelligence. But in today's world things are not so simple. In many cases, surveillance will have two key goals — the gathering of foreign intelligence, and the gathering of evidence for a criminal prosecution. Determining which purpose is the 'primary' purpose of the investigation can be difficult, and will only become more so as we coordinate our intelligence and law enforcement efforts in the war against terror".[106]

The foreign intelligence investigation seeks information about other countries and their citizens, and being a foreigner is enough for

---

[104] *United States v. Pelton* (1986) 696 F.Supp, p. 156.

[105] 50 U.S.C. 1804.

[106] J. P. Bjelopera *et al.*, The Federal Bureau of Investigation and Terrorism Investigation, 7–5700, *Congressional Research Service for Congress*, p. 4.

it to be activated. It is directed at foreign governments, international terrorists, spies, and saboteurs. It attempts to ease some of the restrictions on foreign intelligence-gathering within the US so that there is greater access to information during a criminal investigation, and to establish safeguards against official abuse.[107]

In the past, the maximum duration for FISA surveillance orders and extensions was 90 days and 45 days for physical search orders and extensions, unless they were directed at a foreign power.[108] Section 207 of the Act has now extended the maximum duration of physical search orders to 90 days, and if both surveillance orders and physical search orders are used, extends the maximum life of an order involving an agent of a foreign power to 120 days, with options to extend it for up to a year.[109] Therefore, it gives the prosecutor the right to examine any means of communication which suspected terrorist collaborators use anywhere. Previously, the law granted access to wiretap a specific phone line in a specific location, but it now makes it easier for law enforcement to follow and record a suspect's use of the internet. Otherwise, it would be very difficult to capture and convict terrorists. Such an enhanced surveillance procedure grants law enforcement the freedom to delay notificating a suspect about the search warrant that has been conducted on his/her property. This method can be very useful in prosecuting cyber terrorists' suspects.

It facilitates 'sneak and peek' warrants which "authorizes officers to secretly enter, either physically or virtually; conduct a search, observe, take measurements, conduct examinations, smell, take pictures, copy documents, download or transmit computer files, and the like; and depart without taking any tangible evidence or leaving notice of their presence". In order to apply sneak-and-peek searches, a court must show the required probable cause, as well as finding that there is reasonable cause to believe that immediate notification will have an adverse result.[110]

---

[107] J. W. Whitehead 'Forfeiting "enduring freedom" for "homeland security": A constitutional analysis of the USA patriot act and the justice department's anti-terrorism initiatives', *American Law Review* (2002) 51, pp. 1102–1105.

[108] 50 U.S.C. 1805(e), 1824(d).

[109] 50 U.S.C. 1805(e), 1824(d).

[110] Section 213 Patriot Act.

It is an exception to the 'knock and announce'[111] rule. This warrant provides a good opportunity to search the cyber terrorist's computer without giving them the chance to erase data and modify equipment.[112] This authority is offered by Section 213 of the Patriot Act, but has also been criticised on the grounds of the Fourth Amendment.[113] This section enables law enforcement and intelligence agents to coordinate terrorism investigation without fear of running afoul of the law.[114]

Such acts seem in contrast to the Fourth Amendment which requires officers to knock and announce their purpose before entering to execute a warrant. Nevertheless, it recognises exceptions for exigent situations that may lead to the destruction of evidence, or flight of a suspect. The delayed notice of the installation of an interception device is permitted under the Federal Wiretap Act 1968. Therefore, it must be noted that the Fourth Amendment does not impose demands where it does not apply. Consequently, in permitting the delayed notification of the search of email content in remote storage with a third party for more than 180 days, it does not offend the Fourth Amendment.[115]

The FISA pen register and trap and trace device[116] limited order applies to the facilities used by foreign agents or those engaged in

---

[111] It authorises police officers to delay notification of searches in case that would jeopardise the investigation, although the government can break into a suspect's house without notifying him/her while he/she is not present in the house until after the search had been conducted.

[112] B. A. Shumate, 'From "Sneak And Peek" to "Sneak and Steal"', *Regent University Law Review* (2006) 19, pp. 207–210.

[113] B. A. Shumate, 'From "Sneak and Peek" To Sneak and Steal": Section 213 of The USA Patriot Act', p. 220.

[114] M. A. Baginski, Subcommittee on Crime, Terrorism, and Homeland Security House Committee on the Judiciary (2005) the FBI Federal bureau of Investigation. Available at: http://www.fbi.gov/news/ importance-of-usa-patriot-act-to-fbi-information-sharing (20 May 2013).

[115] *United States v. Miller*, [1976] 425 U.S., p. 435.

[116] The term "trap and trace device" means a device or process which captures the incoming electronic or other impulses which identify the originating number or other dialing, routing, addressing, and signalling information reasonably likely to identify the source of a wire or electronic communication, provided, however, that such information shall not include the contents of any communication. (18 U.S.C. § 3127(3)).

international terrorist or clandestine intelligence activities. Generally, the Pen/Trap statute regulates the collection of addressing and other non-content information for wire and electronic communications. This order is used as part of the investigation to protect against international terrorism or clandestine intelligence activity.[117] It authorises government to apply an order to a court to authorise them to install a pen register or trap and track device in case "the information likely to be obtained is relevant to an ongoing criminal investigation".[118] The FISA Act asks for the court order during foreign intelligence investigation procedures to seize any tangible items regardless of who is in possession of the items. The pen register and trap and trace authority of the FISA permits it to use it to capture the source and destination of information relating to electronic communications and telephone communications.[119]

Section 203 incorporates the Justice Department's Consultation Draft which proposed the expansion of the intelligence community's access to information collected as part of a criminal investigation. It elaborates the duty and procedures of a grand jury. The purpose of the grand jury is to determine if a crime has been committed and if so by whom. The probing of the grand jury may begin without probable cause or any suspicion. Their proceedings are conducted out of court. The matters occurring before the grand jury are completely secret and may only be disclosed in state criminal proceedings or under court-ordered judicial proceedings. Furthermore, the grand jury is allowed disclosure of matters occurring before them to "any federal law enforcement, intelligence, protective, immigration, national defense, or national security" to assist in the performance of their official duties.

B. Search warrant by national security letter (NSL) in the US

The NSL provisions of the Patriot Act (2001) expand the FBI's authority to request customer records from ISPs without prior court

[117] Section 214 Patriot Act (2001).

[118] Section 3122(b)(2) United States Code (Act 18).

[119] Section 214 Patriot Act (2001).

approval.[120] This letter is requested by the FBI or other government agencies from a third party with authority to conduct national security investigations. The NSL has two significant limitations: first, it is only available for authorised national security investigations (international terrorism). Second, it can only be used to seek certain transnational information. Furthermore, it can only acquire the content of any communications. It enables the FBI to obtain a vast amount of information about innocent people such as the websites they visit, lists of email addresses with which a person has corresponded, or the identities of anonymous persons who have posted political speeches on political websites.[121]

Following 11 September 2001, by authorising the NSL through the Patriot Act (2001), the number of NSLs increased. So much so that it appeared that the FBI abused the NSL power, and the American Civil Liberties Union (ACLU) has challenged this in three cases. In *Doe v. Holder*, that was settled in 2010, John Doe could publicly identify himself and his former company as plaintiffs in the case, and part of the NSL was rendered unconstitutional by numerous courts. In 2006, a federal district court ruled that a librarian violated the First Amendment on an NSL which was served on a consortium of librarians in Connecticut. Finally, the government withdrew both the gag and its demand for records.[122]

Demanding this letter does not require any probable cause or judicial oversight. Any person who knows about an NSL in use is given a "gag order" so the persons targeted are not given notice that they are under surveillance. It is especially used in terrorism and espionage investigations. In 2001, Section 505 of the Patriot Act

---

[120] National security letter expanded by the USA Patriot Act under Section 2709 and 505 of the United States Code. The FBI allowed to secretly demanding data by these letters on communications service providers like phone companies and ISPs, about ordinary American citizens' private communications and internet activity without any meaningful oversight or prior judicial review.

[121] W. McCormack, *Legal Responses To Terrorism*, 2nd Edn, Lexis Nexis Publication, United states of America, 2008, pp. 80–84.

[122] American Civil Liberty Union. Available at: http://www.aclu.org/national-security-technology-and-liberty/national-security-letters (27 May 2012).

(2001) greatly expanded the use of the NSL. Also, the Department of Homeland Security has the same ability to use the NSL. However, in 2008, Congress placed new controls on FBI use of the NSL. It tightened the use of this letter with the requirement that they clearly pertain to investigations of a foreign power or an agent, instead of just being considered "relevant" to such investigations.[123]

In 2004, the Federal Court of the US struck down the Patriot Act (2001)'s provision that gave the government authority to issue NSLs without any prior checking to obtain sensitive customer records from ISPs and other businesses without judicial oversight.[124] It was also claimed that some of the provisions were against free speech. In supporting this statement, the ACLU and the New York Civil Liberties Union said that, "This is a landmark victory against the Ashcroft Justice Department's misguided attempt to intrude into the lives of innocent Americans in the name of national security".

i. *The dilemma of protecting national security and privacy of citizens*

As it mentioned in above sections, FISA and the Patriot Act give frighteningly unlimited investigative power to the government in an attempt to making the US a safe place. Section 215 of the Patriot Act[125] gives the FBI unprecedented access to sensitive, personal records and any tangible things. Therefore, an overarching tension is created between the first [126] and fourth constitutional amendments

---

[123] C. Johnson, Lawmaker Wants FBI Access to Data Curbed, Washington Post Staff Writer, United States of America, 2008, p. 10.

[124] C. Johnson, Lawmaker Wants FBI Access to Data Curbed, p. 7.

[125] Section 215 of the U.S. Patriot Act states, "access to records and other items under the Foreign Intelligence Surveillance Act... may make an application for an order requiring the production of any tangible things for an investigation to protect against international terrorism or clandestine intelligence activities, provided that such investigation of a United Sates person is not conducted solely upon the basis of activities protected by the First Amendment to the Constitution".

[126] The First Amendment states "Congress shall make no law respecting an establishment of religion, or exercising the free exercise thereof, or abridging the freedom of speech, or the press; or the right of the people peaceably to assemble, and to petition the government for a redress of grievances".

and the Patriot Act; thereby a complexity is aroused between personal freedom and national safety. In fact, there are supporters and critics of the Patriot Act. Supporters of the Patriot Act believe in the necessity of a nation defending itself in order to provide security for its citizens, while critics of the act, believe that the constitutional rights reserved to individual citizens, cannot be violated even in the name of national security.

The Patriot Act was passed as a result of the September 11 attacks for the sake of US security and to protect the US. Therefore, some aspects of the act created risks for American's civil liberties. Although achieving a proper balance between security and liberty is a high concern of civil societies, the balance can shift to some degree in favour of security in an emergency time like the September 11 incident. However, once such new laws are enacted, they are difficult to change. The Patriot Act is the example of such an enactment in a time of emergency and when a security threat existed. The Patriot Act increases the police power at the domestic and international level. This is due to the fact that it has eliminated many of the checks and balances that allowed the judiciary system to ensure that state police power is not abused.

However, this act has come under strong fire from civil rights movement organisations. These movements believe that this legislation is a threat to individual rights that are protected by the first, fourth, fifth, sixth and fourteenth Amendments to the US Constitution. The statistics indicate that this act was created for a time of emergency and for the time being the priorities of the people have changed and they have discovered that their privacy and civil right are simultaneously important besides security. On one side is the citizen's right to live in a secure environment and on the other side, the right to live without any invasion of their privacy. From 2004, Congress's views also shifted by blocking passing of legislation addressing the September 11 attacks. The current American situation is totally different from the time of when the Patriot Act was passed; US citizens have switched back to their previous model by starting to consider the possible threat to their privacy which was brought on with the passing of the Patriot Act.

The first 10 Amendments to the US Constitution declare certain basic freedoms and procedural safeguards designed to protect citizens from governmental power. It is hard to motivate people to undertake private costs or forgo private benefits for the collective good. It can be said that security brings so many benefits to American society, but it has a cost. People may lose their civil liberties for paying the alleged cost.

4.6.4 Search Warrants in Cyberterrorism Cases: The UK Experience

A 'control order' was introduced by the Anti-Terrorism, Crime and Security Act 2001 in UK. It enabled the Home Secretary to order the detention of asylum-seekers and immigrants who were suspected of involvement in international terrorism for an indefinite period. It could be issued based on the Secretary's approval without requiring a conviction. However, it was only applicable to non-UK nationals. This was redressed in the Prevention of Terrorism Act 2005 by the introduction of a new form of control order which was applicable to UK citizens as well as foreign nationals.[127]

Due to the rapid and widespread public use of the internet, electronic information requires informational privacy, which is mostly provided by digital encryption. Previously, people secured their properties with metal keys. The quality of the content has changed nowadays, and encrypted data is unintelligible if there is no encryption key to decrypt it.[128] According to Member of Parliament Charles Clarke, encryption is a "double-edged sword", because serious criminals such as cyber terrorists use encryption methods to conceal their misdemeanours, while the government has a duty to facilitate state access to the material necessary to decrypt data. Some data uses non-

---

[127] M. Charvat, 'A study of UK anti-terror law', in Centre of Excellence Defence Against Terrorism *Legal Aspects of Combating Terrorism*, Centre of Excellence Defence Against Terrorism, IOS Press, Ankara, Turkey, 2008, pp. 109–110.

[128] B. B. Chatterjee, 'New but not improved: A critical examination of revisions to the regulation of investigatory powers act 2000 encryption provisions', *International Journal of Law and Information Technology* (2011) 19(3), p. 247.

recovery information which severely affects the ability of the government to fight against terrorism. Although it is possible to crack the encryption key, it is very difficult and takes considerable effort, time, and cost. As a result, the Regulation of Investigatory Powers Act (RIPA) was passed in 2000 relating to state access to encryption keys. It is used for suspected terrorist activity occurring in the virtual world in order to protect national security.

Section 49 of this Act empowers authorised agencies to serve notices for disclosure of encrypted material. The recipient of a Section 49 notice becomes guilty of an offence if he fails to make the disclosure required knowingly. If the accused shows that the key is not in his possession and he can provide adequate evidence about this fact, the contrary is not proven. If no defence is raised, the penalties as originally enacted under Section 53(5) are: "(a) on conviction on indictment, to imprisonment for a term not exceeding 2 years or to a fine, or to both; (b) on summary conviction, to imprisonment for a term not exceeding 6 months or to a fine not exceeding the statutory maximum, or to both".[129] If the content of a Section 49 notice contains a secrecy clause which is then disseminated, this could be punished by imprisonment for up to 5 years.[130]

The penalty which was proposed in Section 49 showed that the government placed national security at the highest level and that this justified a higher penalty. Thus, the major clause in the Terrorism Act 2005 reflected an extended detention without trial.[131] Such suspects and crimes deserved to be punished at the most severe level, and the decryption assisted police to identify victims and further suspects. As time is very important in police investigations, early detection and intervention is of paramount importance to fight counter-terrorism. Thus, the measures reflected in the Terrorism Act 2005 that extended detention without charge.[132] In terrorism cases, evidence gathering relies on surveillance agents in building up a tentative case. In such serious offences, the solicitor will advise their clients to remain calm,

[129] RIPA 2000, Section 53 (5)(a)(b).

[130] RIPA 2000, Section 54 (4)(a).

[131] Terrorism Bill 2005, clauses 23 and 24.

[132] Terrorism Bill 2005, clauses 23 and 24.

while the building up of such solid evidence is a pressing matter for the prosecutor.

Therefore, for serious offences the police should buy more time, particularly where the evidence is apparently scant due to encryption and where the potential results could be enormous.[133] Consequently, a Section 49 notice can be issued on the grounds of national security. Section 15 of the Terrorism Act 2006 amended Section 53 RIPA 2000 to comply with Section 49 notices for a case involving national security. The maximum term for failure is 5 years, whereas for any other case, the maximum is 2 years. Section 15(2) of the Terrorism Act 2006 inserted subsections (5A) and (5B) to Section 53(5) RIPA 2000 in order to provide a higher sentence for cases involving national security. A national security case is defined as "a case in which the grounds specified in the notice to which the offence relates as the grounds for imposing a disclosure requirement were or included a belief that the imposition of the requirement was necessary in the interests of national security".[134] This definition can be applied to cyberterrorism cases that involve national security. A cyber attack can threaten the national security of a country.

Section 41 of the Terrorism Act 2005 authorises the police to arrest and detain terrorist suspects for up to 48 hours without any charge. The police can persuade the judge to extend the period of detention for up to 7 days. After that, it may be increased to 14 days by the Criminal Justice Act 2003 and up to 28 days by the Terrorism Act 2006.[135] Under Section 44, police are given the power to stop and search any vehicle or person in certain areas within their jurisdiction without any reasonable suspicion that the vehicle/person that they stop and subject to a search might have a connection with terrorism. However, Section 44 was condemned as illegal by Article 8 of the European Court Of Human Rights as it lacked the

[133] B. B. Chatterjee, 'New but not improved: A critical examination of revisions to the regulation of investigatory powers act 2000 encryption provisions', *International Journal of Law and Information Technology* (2011) 19(3), p. 271.

[134] RIPA 2000, Section 53(5)(5B).

[135] Section 41 Terrorism Act 2006.

adequate legal safeguards against abuse. The UK Commission in 1998 said that they did not have sufficient evidence of the scale and results of computer misuse to reach the point of requiring legislative action.[136]

One of the various forms of cyberterrorism is information warfare. Information warfare involves direct attacks and ignores the principal of rights, which can result in threats to individuals. It might involve defacing a website, use of a virus and a DoS attack. It can even cause aircraft to collide by interfering with their computer systems. Although the offence involves terrorists gaining information by intelligence gathering via the internet, the results of this offence can be detrimental to individual safety and security.

By introducing a 'freezing order' in the Anti-Terrorism, Crime and Security Act 2001 Act, the government was authorised to freeze the assets of persons or organisations suspected of being involved in terrorism. This enabled the Treasury to freeze the assets of overseas governments or residents who had taken action, or had intent to act, against the UK's economy, or any special act which constituted a threat to the life or property of a national or resident of the UK.[137]

Offences in Sections 2 and 3 of the Computer Misuse Act are serious arrestable offences in the Criminal Evidence Act of 1984 related to the granting of search warrants. In such a case, if the arrestable offence was committed and the relevant evidence to the case was found on specified premises, a search warrant may be issued by a justice of the peace. This warrant empowers the seizure of any item of property related to the offence under investigation. For cyberterrorism and cyber crime cases where information is stored in a computer, the prosecutor must printout that information if it is considered "necessary to do so in order to prevent it being concealed, lost, tampered with or destroyed".[138]

---

[136] J. Clough, *Principles of Cyber-crime*, p. 129.

[137] M. Charvat, A study of UK anti-terror law, p. 109.

[138] Section 19(4) Regulation of Investigatory Power Act.

4.6.5 Search Warrants in Cyberterrorism Cases: The Malaysian Scenario

Computer forensics and investigation is a new area in Malaysia. However, crimes must be successfully prosecuted in order to be solved and proper investigation processes will lead to a better prosecution.

However, determining the intention of a perpetrator is ambiguous and vague. "It is difficult to distinguish where the act or threat is intended or may reasonably be regarded as being intended to be done".[139] In doing so, according to Section 106C, the public prosecutor has the power to authorise a police officer to intercept any communication that the public prosecutor suspects is likely to contain information relating to the commission of a terrorism offence. It provides the public prosecutor with unfettered authority to intercept private communications so long as he considers it "likely to contain any information relating to the commission of a terrorism offence". Therefore, in a case of suspected cyberterrorism, any conversation can be intercepted. Put another way, it gives a wide power to the public prosecutor to intercept, listen, and install communications. Furthermore, the information will be admissible at the accused's trial as evidence. Conversely, scholars believe that the power authorising the public prosecutor to intercept the communication should be given to the court.[140]

It appears that this power is too wide. It may be misused by the public prosecutor because it may infringe upon the privacy of the individual; therefore, the prosecutor should be limited by this provision only being exercised to communications pertaining to a person suspected of being involved in a terrorist offence.[141] Moreover, more

---

[139] Available at: http://ejp.icj.org/img/cpc_terrorism_amendment.pdf (22 Feb 2011).

[140] Malaysia, Memorandum to the Special Select Committee on Penal Code (Amendment) and Criminal Procedure Code, Joint Action Group Against Violence Against Women, 28 October 2004, p. 6.

[141] Malaysia, Memorandum to the Special Select Committee on Penal Code (Amendment) and Criminal Procedure Code, Joint Action Group Against Violence Against Women, p. 6.

safeguards should be added to the implementation of this act. Also, the court should have the authority to decide whether evidence thereby acquired is admissible as evidence at the trial of an accused person. Section 272B allows for live video evidence to be admitted in court at the Minister's discretion. Section 107A permits an informant to "request a report on the status of the investigation of the offence complained of".

This provision does not seem to be in harmony with the spirit of the Federal Constitution. According to Article 8(1) of the Malaysian Constitution, "all persons are equal before the law and entitled to the equal protection of the law". The Constitution emphasises the principle of natural justice and governing the country in accordance to the rule of law. The right to privacy is natural and this right of personal liberty is protected by the Federal Constitution. It states that "no person shall be deprived of his life or personal liberty save in accordance with law". It declares that no one is above the law. Therefore, a law that deprives people of their privacy is not in accordance with the spirit of the Federal Constitution, according to Article 8(1).

Suara Rakyat Malaysia (SUARAM) believes that this authority should not be given to public prosecutors, since they are not accountable for their actions and judicial review. SUARAM thinks such authorisation should be given to persons who are accountable and under a duty to justify their decisions. Furthermore, the basis of this provision does not lie on standard ground, due to firstly, the results that have been obtained from the action are not considered by an independent communications commission which has the duty to review the results and they do not have certain procedures which have to be followed by the prosecutor.[142] Consequently, they do not have an independent tribunal to consider the functions of agencies that carry out interception for complaints which they have received.

Secondly, Section 106C states: "if the public prosecutor considers that it is likely to contain any information relating to the commission of a terrorism offence". The application of this provision is problematic,

[142] Available at: http://ejp.icj.org/img/cpc_terrorism_amendment.pdf (22 Feb 2011).

because it does not make it necessary for the public prosecutor to have reasonable grounds or "probable cause" to suspect someone.

Also, under clause 2 of Section 130C, "sufficient evidence" that a terrorist act has been committed is a very ambiguous term. "Sufficient evidence" is stated as a "certificate purporting to be signed by an appropriate authority to the effect that the item or substance described in the certificate is a weapon, a hazardous radioactive or harmful substance, a toxic chemical or a microbial or other biological agent or toxin shall be sufficient evidence of the fact stated in it". However, the appropriate authority is not defined with certainty and it is not known who will be deemed an expert to sign this evidence. There are no guidelines in the provision to clarify what type of material should be given. Therefore, these ambiguous provisions leave room for misuse.[143] Applying this provision to cyberterrorism cases is difficult; and the instances given in the Act are not exclusive and are only examples. It is submitted that the broad definition of the Penal Code in defining terrorists and the inappropriate judicial proceedings would be as dangerous as terrorism itself.

Malaysia has always had a specific Act that deals with terrorism or any form of threat to national security, namely the Internal Security Act 1960 which was drafted in accordance with Article 149 of the Federal Constitution. However, the Internal Security Act 1960 has been repealed and the government has attempted to enact a new rule modelled on the Patriot Act (2001) of the US and the Anti-Terrorism Act in the UK. The Internal Security Act was replaced by the Security Offences Act 2012. It provides special measures relating to security offences.

"Special offences" in clause 3 is defined as offences specified in the First Schedule. There are two categories of security offences listed in the First Schedule, both of which are already found in the Penal Code. The first category is "Offences against the State" which appear in Chapter VI of the Penal Code, that is, from Sections 121 to 130A. The second category is "Offences relating to Terrorism" which can be found in Chapter VIA of the Penal Code, that is, from Sections

---

[143] Available at: http://www.aliran.com (22 Feb 2011).

130B to 130T.[144] Generally, offences against the State in this category share a common theme of force being used or intended to be used in order to obtain their objectives, in order to overthrow the government or to harm members of the different branches of government.

The Security Offences Act offers prosecution of Chapters VI and VIA of the Penal Code in a very oppressive manner compared with the prosecution and trial under the Penal Code. This intention is revealed by a brief review of its provisions. Special powers are conferred on the police for security offences under Part II of this Act. A police officer is empowered to arrest and detain any person "whom he has reason to believe to be involved in security offences" without warrant according to clause 4(1). Under clause 4(5) the person can be detained for a maximum period of 28 days for investigation. Under clauses 4(6) and 7(1), by an Order of Court, an electronic monitoring device may be attached to that person, even after the release of that person. The detainee can be kept without notification to his lawyer for a period of 48 hours.[145]

Part IV of the Act relates to sensitive information. It empowers the court to hear and store such sensitive information on camera. Such information may be admissible as evidence and can be disclosed to the accused under clause 8(7). Such a decision cannot be appealed. "Sensitive information" is defined in clause 3 as:

> Any document, information and material a) relating to the Cabinet, Cabinet committees and State Executive Council; or b) that concerns sovereignty, national security, defence, public order and international relations whether or not classified as 'Secret' or 'Confidential'.

This is a wide definition which can cover every governmental document, information, or material. Part VII of the Act removed all protection that was given to the accused by the Evidence Act 1950. Consequently, under clause 18, statements of persons who are dead

---

[144] Available at: http://www.themalaysianinsider.com/sideviews/article/is-the-security-offences-bill-constitutional-tommy-thomas/ (14 Jul 2012).

[145] Suara Rakyat Malaysia. Available at: https://www.facebook.com/suararakyatmalaysia (22 Aug 2012).

or cannot be found or are incapable of giving evidence shall be admissible as evidence, although it is unfair since they cannot be cross-examined.

In view of these concerns, the Security Offences Act 2012 does not provide full protection as it lacks judicial scrutiny and it provides arbitrary powers to the police to arrest and detain a person without warrant. It gives broad powers to the police by allowing them to detain a person without being brought before a judge and it can be up to 48 hours before the suspect is given access to a lawyer. This violates the fundamental rights and liberty of individuals. The police can detain anyone opposed to them because the law allows them to detain any person on the basis of belief that the person may be involved in a security offence. Such vagueness would give the police extensive opportunity to abuse their powers and perform searches and intercept communications without a judicial warrant. It appears to be a serious infringement of personal liberty to permit the police to impose electronic monitoring devices on a person who is released from detention.[146] It seems that the repeal of the Internal Security Act does not limit the wide powers which are given to the police and that the Security Offences Act once again provides excessive powers to the police with limited oversight or accountability.

Apart from an Act's provisions, implementing the Act and investigating a computer crime are the key ingredients in making the Act effective. Sections 10 and 11 of the Computer Crime Act 1997, give authority to the Malaysian police to investigate computer crime. Consequently, cyberterrorism can be prosecuted under this Act as a part of cyber and computer crime. Thus, an officer above the rank of inspector is empowered under Section 10 of the Act to seize evidence of a cyberterrorism offence in reasonable instances, and they can even arrest a person without a warrant where the police reasonably believe that the accused has committed or is committing an offence under this Act. This applies to both victims of computer crime and perpetrators because the evidence of computer crime might be found on a

[146] Suara Rakyat Malaysia. Available at: https://www.facebook.com/suararakyatmalaysia (22 Aug 2012).

victim's computer and "the police officer has reasonable grounds for believing that by reason of the delay in obtaining a search warrant the object of the search is likely to be frustrated". The term "computer crime" includes a wide range of criminal offenses. This scope may become even wider with the subtitled term "computer-related crime". According to this notion, cyberterrorism can be included as a criminal offence under this act. These provisions can also be applied to cyberterrorism cases as long as no specific legislation about cyber-terrorism exists.

It appears that Section 11 mandates victims to report any computer crimes; if the computer crime is not reported, this could hinder the police officer who is investigating the computer crime. However, in reality, most businesses fail to report being affected by a computer crime, since they fear negative publicity and fear diminished popularity among clients and investors. Nevertheless, a mandate to report computer crime would assist law enforcement to a great extent.[147] Most countries, including Malaysia, have established a Computer Emergency Response Team (CERT) to respond to computer crime incidents and to create a safer cyber space in order to promote national sustainability.[148]

4.6.6 Prosecution in Cyberterrorism Cases: Comparative Analysis between the US and the UK

The RIPA in the UK empowers authorised agencies to serve notices for disclosure of encrypted material. The amended terrorism act extends the detention without trial for the police. As time is very important in police investigations, early detection and intervention is of paramount importance to fight counter-terrorism. Thus, the measures reflected in the Terrorism Act 2005 extended detention without charge.[149] For such serious offences the police should buy more time,

---

[147] C. Yew Wong, Malaysian law and computer crime, *Sans Institute InfoSec Reading Room*. Available at: http://www.sans.org/reading_room/ (16 Nov 2011).

[148] Cyber security Malaysia. Available at: http://www.cybersecurity.my (16 Nov 2011).

[149] Terrorism Bill 2005, clauses 23 and 24.

particularly where the evidence is apparently scant due to encryption and where the potential results could be enormous and where the case could involve national security.[150] However, the Detention without Trial of Terrorism Act in the UK was condemned as illegal by Article 8 of the European Court of Human Rights because it lacked the adequate legal safeguards against abuse. The law can be extended without sufficient reason because of the seriousness of the offence. In computer-related offences such as cyberterrorism, data is fragile and sensitive, and search warrants can prevent the removal of data. For a search warrant to be issued in the US, a federal agent must prove that under the circumstances known to him or her, there is a reasonable belief that a person has committed, is committing, or is about to commit a crime based on the Fourth Amendment of the US Constitution.

According to the ECPA it is illegal for police or the government to use a tapped phone unless they have obtained a search warrant. The citizen's consent must be obtained. It seems that the US in contrast with other nations provides a comprehensive tool for search warrants. Simultaneously, this comprehensive regulation prevents government circumventing their position in cyberterrorism cases from handing over private conversations of people without warrant. New amendments in the Patriot Act make it easier for law enforcement officers to investigate computer crime. It grants federal officials greater powers to trace and intercept terrorists' communications. This law is the main national counter-terrorism tool that expands the government's ability to monitor Americans in the name of national security. Through the search warrant, an investigator can obtain any personal information case that may involve cyberterrorism.

The accessibility of this information operates on the principle of "necessity"; it prohibits the seizure of any tangible property or electronic communication except when the court finds it a reasonable necessity.[151] Furthermore, the US has two more statutes under the Patriot Act for investigation. According to the NSL provisions of the

---

[150] B. B. Chatterjee, 'New but not improved: A critical examination of revisions to the regulation of investigatory powers act 2000 encryption provisions', *International Journal of Law and Information Technology* (2011) 19(3), p. 271.
[151] 18 U.S.C. 2703(c).

Patriot Act 2001 the FBI's authority can be used to request consumer records from ISPs without prior court approval for national security investigations. Applying for a surveillance order under the other statute known as FISA requires a certification that "the purpose for the surveillance is to obtain foreign intelligence information".[152]

The application for FISA must "contain a certification by a designated official of the executive branch that the purpose of the surveillance is to acquire foreign intelligence information". In fact, if the primary purpose of FISA is not for obtaining foreign intelligence information, it would not be offered. Under the new amendment of foreign intelligence investigation in the Patriot Act (2001), foreign intelligence needs only be "a significant purpose of the surveillance". It expands the definition from agents of a "foreign power" to persons associated with "a group engaged in international terrorism or activities in preparation".

A 'sneak and peek' warrant facilitates the job of the prosecutor. It "authorizes officers to secretly enter, either physically or virtually; conduct a search, observe, take measurement, conduct examinations, smell, take pictures, copy documents, download or transmit computer files, and the like; and depart without taking any tangible evidence or leaving notice of their presence". In order to apply sneak-and-peek searches, a court must show the required probable cause, as well as finding that there is reasonable cause to believe that immediate notification will have an adverse result.[153] It provides a good opportunity to search the cyber terrorist's computer without giving them the chance to erase data and equipment.[154] Nevertheless, it recognises exceptions for exigent situations that may lead to the destruction of evidence, or flight of a suspect.

The regulation of search warrants in Malaysia grants this search warrant to the police by the prosecutor as in the UK. It provides the public prosecutor with unfettered authority to intercept private communications so long as he considers it "likely to contain any

[152] 50 U.S.C. 1 804(a)(7)(B).

[153] Section 213 of Patriot Act 2001.

[154] B. A. Shumate, 'From "sneak and peek" to "sneak and steal"', *Regent University Law Review* (2006) 19, pp. 207–210.

information relating to the commission of a terrorism offence". Therefore, in a case of suspected cyberterrorism, any conversation can be intercepted. Put another way, it gives a wide power to the public prosecutor to intercept, listen, and install communications.

To sum up, all the provisions of search warrants in both regions (US and UK) following the September 11 attacks extend their search warrant duration. It is beneficial for government and it gives the prosecutor the right to examine any means of communication which suspected terrorist collaborators use anywhere. It allows law enforcement to follow and record a suspect's use of the internet. However, it seems that the privacy of people is infringed in some aspects.

To obtain a search warrant, the requesting party or investigator must provide sufficient cause that a crime has been committed and that the electronic device located at the place was involved in committing the crime. Then the investigator must obtain a warrant from the court for search, seizure, or arrest of suspect. However, the search must be within the scope of the warrant. The digital evidence resulting from the search warrant enables the prosecution to convict criminals.[155] Search warrants are applied in two ways: by obtaining judicial authority before searching and seizing or vesting this authority to prosecutors or police. An overview of the laws relating to search warrants in three jurisdictions (the US, the UK, and Malaysia) indicates that detention without trial occurs in all regions. Given that time is very important in police investigations, early detection and intervention is of paramount importance to fight counter-terrorism. As such, detention without charge can be extended, although it may be misused by public prosecutors because this statute may infringe upon the privacy of the individual. Public prosecutors are not accountable for their actions and judicial review.

It seems that such authorisation should be given to persons who are accountable and under a duty to justify their decisions. The results that have been obtained from the prosecutor's actions are not considered by an independent communications commission which has the

---

[155] A. Khan *et al.*, Digital forensics and crime investigation: Legal issues in prosecution at national level, *IEEE Computer Society*, p. 244.

duty to review the results and they do not have certain procedures which have to be followed by the prosecutor.[156]

4.7 EXTRADITION

The whole process of investigation clarifies that the existence of mutual legal assistance (MLAT), and bilateral and multilateral treaties among countries is vital for the capability of countries to prosecute cyber attacks. The absence of an extradition treaty between two countries which renders the prosecution of the attacked country. (The country who received the attack cannot prosecute the attacker party, in case there is no extradition treaty between them. Serious crime is increasingly becoming international and thereby criminals can flee from justice by crossing borders. In order to tackle this development a judicial cooperation between nations is needed. Developed countries reformed their extradition law to contribute to that process.[157] For the purpose of the extradition, the world is divided into two categories: category one is the Framework Decision on The European Arrest Warrant that create a fast-tracking extradition arrangement with member states of the EU. It covers 'computer related crime', but it can be issued for offences carrying a maximum penalty of 12 months or more imprisonment. Extradition treaty between two countries plays a significant role in prosecuting cyberterrorism offences. Whenever a cyber terrorist case occur between two countries, the right of demanding extradition from a perpetrator's country only exists in the case that is created by treaty.

As discussed previously, the response of prosecuting an attack is based on the initialisation of the attack, whether it is from a private citizen-hacker or a government. In a case that derived from an American citizen-hacker in the US, he was prosecuted for the knowing transmission of a program that caused damage under different

---

[156] Available at: http://ejp.icj.org/img/cpc_terrorism_amendment.pdf (22 Feb 2011).

[157] L. J. Lioyd, *Information Technology Law*, 5th Edn, Oxford University Press, New York, 2008, pp. 272–273.

provisions of the CFAA. In this case, the hacker was aware of the transmission of the program to a protected computer, and that it would intentionally cause damage. He received a maximum penalty of 5 years' imprisonment under Section 1030(a)(5)(A). However, due to the fact that the US does not have any extradition treaty with China, this did not apply for this case. In the cases named Code Red i and Code Red ii which happened in 2001, 359,000 US computers were infected within just 14 hours.

It had two different waves. The first wave caused the Pentagon to temporarily block public access to vital websites and caused a change in the numerical internet address of the White house.[158] It was a time limitation program. A program was installed on computers against various websites to only operate on the 20th–27th days of every month and to perform pre-programmed DDoS network attacks.[159] After the attacks, the websites showed the message "Hacked by Chinese!" upon opening. The Code Red worm attack originated at the Chinese University in Guangdong, China. One of the main targets of the Code Red attack was the US President's White House website. Code Red activated itself at 8 p.m., 31 July 2001, but US Presidential cyber security experts could frustrate the attack. They blocked internet traffic to the White House at the server to prevent the DDoS attack. The Code Red attacks were parallel to the attacks on the Estonian and Georgian presidential websites.[160]

Although both countries have computer crime acts — the US has the CFAA and China added computer crime to the P.R.C. Criminal Code in 1994 — none of them could force the other country — in a case a private citizen-hacker attacking a non-vital target — to prosecute the hacker under its own law, nor could it force an extradition of

---

[158] D. M. Creekman, 'A helpless America? An examination of the legal options available to the United States in response to varying types of cyber-attacks from China', *The American University International Law Review* (2003) 17(3), p. 664.

[159] David Moore and Colleen Shannon, 2008. The Spread of the Code Red Worm, CAIDA: The Cooperative Association for Internet Data Analysis. Available at: http://www.caida.org/research/security/code-red/coderedv2_analysis.xml (22 Apr 2011).

[160] J. F. Dunnigan, *The Next War Zone*. Citadel Press, Charleston, SC, 2003, p. 80.

the citizen-hacker. Generally, if a computer attack is attributed to a private citizen and the person is not connected to any state, the attack is considered as a criminal matter regardless of whether it is a vital target or of the extent of damage, and it will be punished the same as that of an attack on a non-vital target by a US citizen under the CFAA.

There was no extradition treaty between the two countries and even if there was, an extradition treaty problem still remained since most extradition treaties exempted political offences from their content. They would not confer extradition and in this case China considered the citizen-hacker's action as a critique of the American capitalistic regime and a political offence. Upon defacing the website by this worm, the website opened with the words "Hacked by Chinese!". Therefore, if a private citizen was responsible for committing the act, international law was implicated. If the Chinese government was committing the act, it would be controlled by international law.

However, it is impossible to bring a Chinese citizen to the US for justice because there is no extradition treaty between the two countries. In this attack, since one state intentionally attacked the other nation within the territorial boundaries of the other nation, the international law of conflict management was invoked, under the use of force during peacetime article, according to the UN Charter which governed the case of force.[161]

According to this article in the UN Charter, it "prohibits the threat or use of force against the territorial integrity of another nation, unless it is conducted pursuant to a nation's right to self-defence or authorised by the U.N. Security Council".[162] If a non-vital US target suffers from a cyber attack but it causes minimal physical destruction, it will be considered an illegal use of force. The criteria for this kind attack would not qualify it as an armed attack. Minimal damage can be considered as unlawful use of force, but does not qualify as an

[161] D. M. Creekman, 'A helpless America? An examination of the legal options available to the United States in response to varying types of cyber-attacks from China', p. 665.

[162] United Nations Charter, Article 2(4).

armed attack. Thus, the US is not authorised to use the right of self-defence.[163]

Furthermore, claiming self-defence as a state's right is controlled "by the principles of necessity and proportionality, among others, and is prohibited for retaliatory or punitive purposes".[164] Moreover, if the US responds in kind, for instance it distributes similar viruses in China, then it could be viewed as a "retaliatory" or "punitive" use of force (the use of force for retaliatory or punitive actions is prohibited by customary international law). The US could appeal to the UN Security Council under Article 39 of its Charter, because it can respond with force to any incident that threatens peace even if the incident does not have sufficient criteria to qualify as an armed attack. This is because in this kind of attack, if the Security Council determines that the acts have the potential of being a threat against international peace, the Security Council can authorise a proper response. And the UN would provide sanctions if it cannot maintain peace.[165]

There are limited responses available for the US; it can ask for reparation of any damage that it has received and can disclose publicly the Chinese government's role in threatening and carrying out attacks via computers, and this would lead to international embarrassment for China. In conclusion, the US' legality of action against China is still unclear and questionable.[166]

4.8 CONCLUSION

The extreme and unique nature of cyberterrorism cases has turned its enforcement task into a daunting one. The fragile nature of digital evidence and its limitations make them difficult for investigation and

[163] D. M. Creekman, 'A helpless America? An examination of the legal options available to the United States in response to varying types of cyber-attacks from china', p. 667.
[164] D. M. Creekman, 'A helpless America? An examination of the legal options available to the United States in response to varying types of cyber-attacks from China', pp. 667–668.
[165] United Nations Charter, Articles 39, 41, 42.
[166] D. M. Creekman, 'A helpless America? An examination of the legal options available to the United States in response to varying types of cyber-attacks from china', p. 669.

prosecution. Indeed it is a different game altogether, offering huge challenges to both investigators and prosecutors. This chapter concludes by emphasising on the urgent importance of analysing whatever gaps and lacunas there are and rectifying them. Any gap or lacuna should be attended to urgently and wisely in response to the aggressive nature of the cyber terrorists and the immediate danger that they pose to any country. Any country attempting to rectify lacunas and gaps in investigation and prosecution procedures in cyberterrorism cases must enact relevant laws or legal provisions. Improving on the present investigation and prosecution process will significantly increase the chance of bringing these criminals to justice.

CHAPTER V

ISSUES OF JURISDICTION FOR CYBERTERRORISM

5.1 INTRODUCTION

Recent progress and development in computer technology has provided new opportunities for those who are willing to involve themselves in illegal activity and this has thereby created some new varieties of criminal activity that pose challenges for legal systems as well as for law enforcement.[1] Due to the fact that cyber terrorist attacks are conducted in multiple states, the procedure of prosecution is difficult; therefore, the attacked country will invoke international law to seek justice for damage caused. Although countries implement technical measures, legal measures must also be taken in order to prevent and deter the rapid growth of cyberterrorism. Nations must come up with self-regulatory legal mechanisms to combat against the misuse of new technologies; however, such mechanisms need to be supported by international agreements and appropriate national legislation.[2]

The main issue regarding jurisdiction in the international space of the internet is the dichotomy which exists among three components

[1] S. W. Brenner, 'Cyber-crime investigation and prosecution: The role of penal and procedural law', *Murdoch University Electronic Journal Law* (2001) 8(1), p. 1.
[2] L. Bantekas, *International Criminal Law*, 3rd Edn, Routledge-Cavendish Publication, United Kingdom, 2007, p. 265.

of jurisdiction in cyber space that are personal jurisdiction, territorial jurisdiction, and universal jurisdiction. Although many steps have been taken to combat cyberterrorism, from legal to technical steps, these attempts have not been sufficient to prevent cyberterrorism. It appears that greater international cooperation is needed. For the time being, as cyberterrorism cannot be prevented, effective prosecution is a logical method to deter cyber terrorists. International law prescribes several types of jurisdictions: nationality jurisdiction (active personality, passive personality), territorial jurisdiction (objective, subjective), universal jurisdiction, and protective jurisdiction. Of these, territorial jurisdiction and universal jurisdiction are more attuned to deter cyberterrorism. However, universal jurisdiction is the more suited to deter cyberterrorism due to the nature of the internet and the initial reality of cyberterrorism. This is because it ignores national borders.

5.1.1 Objective of the Chapter

Borders are currently being altered within the internet through multiple jurisdictions that have the potential to overlap and cause conflict. There is still much debate over the proper jurisdiction for cyberterrorism. This chapter aims to determine the most suitable jurisdiction for cyberterrorism to solve specific offences related to enforcement (i.e. court's power, investigation, and prosecutions).

5.2 JURISDICTION

International law defines 'jurisdiction' as: "the limits of the legal competence of a State … to make, apply, and enforce rules of conduct upon persons. It concerns essentially the extent of each state's right to regulate conduct or the consequences of events".[3] Jurisdiction refers to the sovereign authority of a nation within its territory to propose legislative, executive, and judicial principles. However, one of the main problems of the jurisdiction attribute is the lack of information

---

[3]V. Lowe, '*Jurisdiction*' *in International Law*, Malcolm D. Evans (Ed.), 2nd Edn, 2006, p. 335.

and absolute certainty. The nature of the internet gives the ability to the user to disguise his identity, leading to inherent difficulties in determining the states that fail to prevent an attack from originating within their borders. Therefore, states must cooperate with each other to share information in order to identify attackers.[4] In the UK, the principle of traditional rules applies for its jurisdiction.

Although there have been several efforts by various scholars around the world, we have yet to attain a uniform response to address the issue of jurisdiction in cyberterrorism. Different legal systems have responded in different ways based upon their own ideas of justice and interest. For instance, the US has exercised personal jurisdiction over its forum, and over foreign defendants, apropose effects doctrine jurisdiction. It admits jurisdiction to the US if an extraterritorial behaviour or crime affects or harms citizens within the US.[5]

5.3 THE EXERCISE OF UNIVERSAL JURISDICTION BY THE INTERNATIONAL COMMUNITY AND STATES AGAINST CYBERTERRORISM

The International Criminal Court (ICC) is an international tribunal that was established as a result of a multilateral treaty devoted to international jurisdiction. Such international tribunals are self-contained systems. However, the constitutive instrument of an international tribunal limits its jurisdictional power. The ICC enjoys an inherent jurisdiction, similar to other international tribunals.[6,7] However, the ICC's jurisdiction is premised on the basis of complementary organs,

---

[4] L. Grosswald, 'Cyber-attack attribution under article 51 of the UN charter', *Brooklyn Journal of International Law* (2011) 36. Available at: http://lexisnexis.com (14 Feb 2012).

[5] A. M. Sachdeva, 'International jurisdiction in cyber space: A comparative perspective', *Computer and Telecommunication Law Review Oxford* (2007) 13, p. 245.

[6] The ICC has its own constitution for sentencing the offenses. It is the world's first permanent court that has jurisdiction over individuals accused of some of the most serious international crimes.

[7] I. Bantekas *et al.*, *International Criminal Law*, 2nd Edn, Cavendish Publishing, United States, 2003, pp. 162–164.

i.e. national courts in cases where states override its authority or where it is unable to carry out investigation or prosecution.[8] A state automatically accepts that the ICC has jurisdiction over four groups of crimes, i.e. the crime of genocide, crimes against humanity, war crimes, and the crime of aggression.

5.3.1 The Exercise of Universal Jurisdiction by the International Community

International tribunals have been created for exercising universal jurisdiction over international crimes. The Rome Treaty was the first treaty that articulated all jurisdictions (jurisdiction to prescribe, adjudicate, and enforce) in one instrument. Using universal jurisdiction to deal with international crime dates back to the 19th century. The anti-slave trading treaty and the Nuremberg tribunals are instances of the efforts of the international community in exercising universal jurisdiction over international crimes.[9]

Generally, the theory of universal jurisdiction, which was extended in the Rome Treaty, derives from the idea that the criminal activity must reach a certain level of harm, or threaten the interest of international society for it to be necessary for all states to apply their laws. The theory of universal jurisdiction permits the international community in prescribed jurisdictions to displace the national law with international law. However, the statutes of the Rome Treaty do not focus exactly on the issue of courts' jurisdiction and leave it up to the complementarity principle and the state consent regime. The court exercises its jurisdiction only in cases "involving the most serious crimes of concern to the international community as a whole".[10]

The jurisdiction to adjudicate subjects criminal defendants to the process of the ICC. The ICC jurisdiction is divested in three ways: firstly, the state's consent excludes some cases from the court's judicial jurisdiction. Secondly, the complementarity principle removes certain

[8] ICC statute, Article 17(1) (a), (b), 2(a).

[9] W. B. White, 'Regionalization of international criminal law enforcement: A preliminary exploration', *Texas International Law Journal* (2003) 38, pp. 730–735.

[10] Rome Statute of The International Criminal Court, Article 5(1).

cases because of prudential concerns. Thirdly, the principle of *ne bis in idem* ('not twice in the same') removes the case from the court if there is the chance of legal action being instituted twice for the same cause of action. The court is qualified to assert jurisdiction only in cases where states and the Security Council lodge complaints.[11]

The jurisdiction of the ICC is not limited to when the Security Council refers a case to the court because there has been a threat to world peace and security, even when the state is not party to the court's statutes. In addition, a case can be referred by the prosecutor or state with the state's consent. Thus, the court receives the jurisdiction "by state parties so long as either the territorial state or the state of accused's nationality is either a party to the statute or has accepted the jurisdiction of the court".[12] However, the case may be that the Security Council receives the jurisdiction in times of crises that threaten peace and security and this may conflict with the state's jurisdiction and in limited circumstances the state may withhold its consent and prevent the ICC from exercising its jurisdiction. In a complementary situation that the court may exercise jurisdiction when the state is unable or unwilling to assert jurisdiction.[13] Offences such as genocide and terrorism that are subject to universal jurisdiction by a treaty are also subject to the perpetrator violating the statutes of the treaty not only to the jurisdiction of all member states of the treaty, but also to the jurisdiction of the treaty organisation itself.[14] Although the universality principle allows states to exercise jurisdiction to enforce their criminal laws through their courts in order to punish universal crime, the ICC simultaneously has jurisdiction, but the latter has primacy over national courts in adjudicating.[15]

---

[11] Rome Statute of The International Criminal Court, Articles 17 and 20.

[12] L. Nadya Sadat, 'Universal jurisdiction: Myths, realities, and prospects: Redefining universal jurisdiction', p. 241.

[13] L. Nadya Sadat, 'Universal jurisdiction: Myths, realities, and prospects: Redefining universal jurisdiction', p. 241.

[14] S. Wilske *et al.*, 'International jurisdiction in cyberspace: Which states may regulate the internet', *Federal Communication Law Journal* (1998) 50, p. 123.

[15] S. Wilske *et al.*, 'International jurisdiction in cyberspace: Which states may regulate the internet', pp. 170–171.

Jurisdiction to enforce is the weakest component of jurisdiction principles. However, the conflict of jurisdictions often happens in executive jurisdiction and the limitations of international law are clearer than the area of legislative and judicial jurisdiction. The efficacy of the ICC in enforcing international criminal law is undermined, since it has no police force.[16]

Although it has been previously mentioned that the territorial principle of jurisdiction is applicable in the UK, it should be said that it is a general jurisdiction. According to Sections 62 and 63 of the Terrorism Act 2000, and Section 17 of the Terrorism Act 2006, the UK has asserted universal jurisdiction over the commission of all types of terrorism. It is stated that: "anyone who commits any types of terrorism offences anywhere in the world should be dealt with under the relevant laws of the United Kingdom".[17] Regarding the general jurisdiction of the UK, it enjoys full jurisdiction over any crime committed on its territory. All lists of offences in Sections 62 and 63 of the Terrorism Act 2000 and Section 17 of the Terrorism Act 2006 will be considered to be committed on UK territory, even when they are committed elsewhere.[18]

However, the assertion of universal jurisdiction on these acts on the basis of customary international law and international treaties is not plausible. The assertion of universal jurisdiction in the UK goes beyond that which the UK is obligated or permitted to claim. It is implausible to say that all or most of the terrorism offences in the world may be characterised as affecting the vital interests of the UK. Even the combination of the nationality principle, passive personality, and the protective principle on the basis of orthodoxy cannot establish a customary basis for the UK's assertion of universal jurisdiction.[19]

Furthermore, the possible grounds for the assertion of universal jurisdiction cannot be found in the jurisdiction available to the UK

---

[16] United Nations, *Handbook on Criminal Justice on Responses to Terrorism*, United Nations Publication, New York, 2009, pp. 63–65.

[17] S. Sibbel, 'Universal jurisdiction and the Terrorism Act', *Cambridge Student Law Review* (2007) 3, p. 13.

[18] Terrorist bombing, terrorist finance.

[19] S. Sibbel, 'Universal jurisdiction and the Terrorism Act', p. 16.

following its ratification of the International Convention for the Suppression of Terrorist Bombing and the International Convention for the Suppression of the Financing of Terrorism. These conventions recognise state parties' jurisdiction on territorial, national, and protective grounds. The treaties provide a treaty-based jurisdiction based on the rule of *aut dedere aut judicare* which "states that where an alleged offender is apprehended in the territory of one of the treaty parties, that state must either extradite the offender to another treaty-party with jurisdiction, or itself begin prosecution".[20] However, the universal jurisdiction asserted in Sections 62 and 63 of the Terrorism Act 2000 and Section 17 of the Terrorism Act 2006 extends beyond *aut dedere aut judicare* in two important ways. Firstly, the scope of the UK's jurisdiction is asserted over a wider jurisdiction, rather than confined to those states that have ratified the conventions. Secondly, when exercising the asserted jurisdiction in Sections 62 and 63, at the time of the assertion, the offender does not have to be within the territory of the asserting state. Thus, under Sections 62 and 63, the UK authorities can issue an arrest warrant for offenders anywhere in the world, besides cooperating with other states to bring about the detention and extradition of these offenders to the UK to face trial.

5.3.2 The Exercise of Universal Jurisdiction by States

States asserting universal jurisdiction over criminals use specifically-adapted internal enactments. Until recently, there have been very few state prosecutions of the types of crimes listed in the Rome Statute. A few examples are the French war crimes trials of Barbie, Touvier, and Papon, Israel's quest of Eichmann, Canada's trial of Finta, and Spain's search for General Pinochet. In each of these cases, each state's national court, although applying national law, was also to some extent applying international law, and in the process, challenging questions were raised on both the substantive law itself and on the procedural system that accompanied it. Although prosecutions in the three types of jurisdictions — prescriptive, adjudicative, and

[20] S. Sibbel, 'Universal jurisdiction and the Terrorism Act', p. 18.

enforcement — share certain similarities in national and international ambits when applied to international crimes, there is one important difference, i.e. that although national courts apply prescriptive norms that apply internationally, their adjudicative powers and authority to enforce them are limited to their own territorial spaces.[21]

It has been postulated that, having a theory of "absolute" universal jurisdiction, i.e. universal jurisdiction not subject to any limitations arising from practical concerns, is appropriate in the current scenario of a few states prosecuting non-nationals for criminal acts. However, if states do enact legislation to punish international criminals, and thus assert jurisdiction over such perpetrators, there should be a case for international law establishing rules to resolve otherwise difficult conflicts of jurisdiction. Since, the conflict arises between the states that established legislations over the same act. This point affects both the application of substantive law and its procedural regime, and is discussed below.

The exercising of universal jurisdiction in cyberterrorism cases by states, in the same way as they treat other international crimes, is characterised by two factors. The first one is the prescriptive norms which postulate that all states apply international norms through their national laws. The problem here is that this statement may be true only in theory, while it could vary considerably in practice. It must not be forgotten that the Rome Statute, by the insertion of Article 10 into the text, called on states to 'improve' international norms in the ICC Treaty by encouraging the development of customary international law beyond the treaty definitions of crimes listed in the Rome Statute. This means that the Statute encourages states to modify the definitions of the Rome Statute through their own specific legislation. Thus, it is completely reasonable to suppose that a state's definition of its ambit varies from state to state. In application, overlap between national legislation and such legislation is not unexpected. The application of universal jurisdiction by the international community causes

[21] M. C. Bassiouni, *International Criminal Law: Multilateral and Bilateral Enforcement Mechanisms*, 3rd Edn, Martinus Nijhoff Publishers, Netherlands, 2008, pp. 207–209.

fewer problems[22] because the international community confers specific rules and definitions, while individual states provide various definitions and rules according to their own legislation.

The second factor is regarding the procedural regimes of substantive law. Even if substantive law norms are constant from state to state, the procedural regimes to which they are subjected may vary considerably. The procedural law, especially in criminal procedure, is almost exclusively local in character. And this also applies to aspects of procedure which are covered by international law: there is very little formal congruence between national and international proceedings. While there have been assertions that national prosecutions must have certain specific rules, there has been almost no integration of national and international legal criminal law systems that prove those assertions correct. Thus, it is not clear which law applies. For instance, whenever one of the procedural laws is in issue, such as immunity granted by municipal law to a potential criminal defendant, it is not clear whether the forum (court) state looks to its own law, the law of the state granting the defendant immunity, the law of the state of the defendant's nationality, the law of the state upon whose territory the crimes were committed (the territorial state), or international law to resolve the problem. This is because public international law has not established a conflict of laws system in such a situation, and on the basis of the Lotus paradigm, every state may apply its law as an independent sovereign unless there is some rule prohibiting it from doing so.[23]

In such an instance, it would be an inconsistency if the forum state applies the law of the state granting immunity as the benchmark for its own exercise of universal jurisdiction. This is because of the uncertain nature of the immunity. For example, if the criminal act is committed as part of an internal conflict by the regime in power, the state granting the immunity will be the state of the defendant's nationality as well as the territorial state. Or, the immunity may have been granted by the regime to itself just before it relinquishes power,

---

[22] G. Bottini, 'Universal jurisdiction after the creation of the international criminal court', *International Law and Politics* (2004) 36, pp. 557–560.

[23] M. C. Bassiouni, *International Criminal Law: Multilateral and Bilateral Enforcement Mechanisms*, pp. 207–209.

or it may be extorted with threats of violence from a succeeding regime. Thus, if it is the law of the forum state that applies to the question of whether such immunity is valid, the choice should be between the law of the forum and international law.[24]

There is general consensus that substantive norms, whether established by treaty or custom, are well-established norms of customary international law, and are, in addition, non-derogable and peremptory or *jus cogens* norms. This consensus was confirmed at the Rome Diplomatic Conference to establish the ICC, where most states approved the codifying of these norms and then universally applying them in instances where the UN Security Council referred a particular case to the ICC. Therefore, a state investigating a non-citizen involved in these types of crimes in an exercise of universal jurisdiction is applying international law, *albeit* through the medium of its national law. It is in doubt, however, if the state is also required in the absence of specific treaty obligations, to apply international rules related to the substantive norm, as there is very little evidence that a state is required to do so.[25]

Finally, it can be said that the conundrum posited by the application of international law by national legal systems has existed for a long time. Although this problem has arisen fairly recently in the international arena, all legal systems determining cases in multiple and overlapping courts have encountered this problem. To see the way forward, a brief look at the US Supreme Court's complex doctrine governing the application of state law by Federal Courts may be instructive. In the case of *Erie Railroad v. Tompkins*[26] the court was faced with the question of which law governed the case, i.e. the Federal law or the state law.

Through a series of complex judgments, the Supreme Court stated that many factors governed the question of whether state or federal law applied. Important factors would be whether the application of state or federal law would be 'outcome determinative', or whether the application of either law was affected by the rights and

[24] L. N. Sadat, 'Redefining Universal Jurisdiction', pp. 247–248.
[25] L. N. Sadat, 'Redefining Universal Jurisdiction', pp. 247–248.
[26] [1938] U.S. 64, p. 304.

obligations created under the applicable state law. Basically, if the state law question was 'substantive', then state law applied; if it was simply 'procedural', then federal law applied.[27]

Furthermore, it is noted that the US Constitution has achieved a balance between federal and state courts. Thus, it is submitted that it may be instructive to consider case law in the multiple and conflicting applications of the law by courts with concurrent jurisdiction elaborated in well-developed legal systems such as the US, as a guide to establishing a doctrine that might ultimately be beneficial to international law and in supporting the maturing of international legal systems.

5.4 CONFLICT OF JURISDICTION

In actual fact, conflict of jurisdictions in cyber space may easily occur. It may occur particularly because the effect of cyberterrorism often takes place in a country or countries other than the country in which the attack originated.[28] A new idea arises here, that since the state has the state responsibility, in order to determine which state has the proper jurisdiction to take action in the ambit of conflict of jurisdiction, territorial jurisdiction is the most feasible jurisdiction to be prescribed. Due to the cross-border nature of cyberterrorism, jurisdiction conflicts may easily occur, because, the effect and start of such crime frequently happens in more than one country. Furthermore, as a specific and holistic jurisdiction and method has not been determined for cyberterrorism in cyber space, the conflict of jurisdictions is not a surprising issue.

In fact, universal jurisdiction is offered as the most appropriate jurisdiction by international and multilateral treaties. The relevant international treaties encourage their member states to expand jurisdiction over international offences. Then such jurisdiction is established with respect to incorporation of municipal law regarding the

---

[27] L. N. Sadat, 'Redefining Universal Jurisdiction', pp. 247–248.

[28] S. W. Brenner *et al.*, 'Approaches to Cyber-crime', *Journal of High Technology Law* (2004) IV(1), p. 40.

international offence. As is articulated in Article 5 of the 1984 UN 'Torture Convention', if the alleged offender is located in a state that does not wish to initiate criminal proceedings, it is obliged to extradite the offender to the country which has the closest connection to the offence. Such extradition is based on a bilateral extradition treaty. The extradition process in universal jurisdiction must be based on the legitimacy of the requesting country. In other words, conflicting extradition requests can be decided on the basis of relevant connecting factors. Furthermore, they must not conflict with other agreed rules of international law.[29]

The issue of jurisdiction conflicts is divided into two categories: negative conflicts and positive conflicts. The former occurs in a situation in which no country claims jurisdiction over a cyberterrorism. If an attack targets a nation via hacking tools and denial of service, the attacked country is qualified to assert jurisdiction on the basis of the location of the computer, the effect of the crime and the nationality of the perpetrator. Or, if the cyber attack occurs via a virus or a targeted content-related offence occurs in one place but simultaneously, numerous other places are involved in launching the attack. Negative jurisdiction occurs if the perpetrator launches a cyber attack from one country which is a safe haven and he is also a national of that country.

A good example of an international treaty here is the Convention of Cybercrime which states in Article 5 that in the case that a target's victims are located in several states, several parties assert jurisdiction over the crime. The Convention states that they must consult with each other to determine the appropriate location for prosecution.[30] Some of the aspects of territorial jurisdiction seem appropriate to settle the conflict that arises among jurisdictions.

Another conflict which must not be forgotten is the positive conflict that happens mostly in cyber space cases, particularly cyberterrorism incidents, since the cross-border nature of these lead to it involving a large number of nations. For instance, the "Love Bug"

---

[29] I. Bantekas *et al.*, *International Criminal Law*, pp. 162–164.

[30] A. A. Cottim, 'Cybercrime, cyber terrorism and Jurisdiction: An analysis of Article 22 of the COE convention on cybercrime', *European Journal of Legal Studies* (2010) 2(3). Available at: http://hdl.handle.net/1814/15118 (29 Mar 2012).

virus or the "Blast Worm" allowed many countries to claim jurisdiction on the basis that the effects were taking place on their territories. For example, when a Polish citizen uses a computer in the Netherlands to hack a Malaysian computer and the data is transferred via Singapore and the US, all these states will be able to claim jurisdiction. Thus, in this situation more than one country can claim jurisdiction over a perpetrator based on the same general course of conduct.[31] However, some circumstances may mitigate the ability of claiming jurisdiction, such as lesser damage compared to that occurring in other involved countries, and the fact of data merely passing through the territory of a country without causing damage.

Although there are some factors in prioritising a jurisdictional claim to resolve and prevent jurisdictional conflict, such as place of commission of the crime, custody of the perpetrator, the amount of harm, the nationality (victim's nationality, perpetrator's nationality), the strength of the case against perpetrator, fairness (anticipated fairness in a specific country), and convenience (the extent that the prosecution would be convenient in a country), conflict still exists in cyberterrorism and cyber crime situations, since every individual factor has its intrinsic problem.[32] Furthermore, such factors would be considered in resolving a positive jurisdictional conflict. Each of these factors must be satisfied in order to establish jurisdictional priorities. However, some factor may seem null. These factors determine the reasonable country that is appropriate to prosecute cyber criminal.[33] Factors do not have equal weight. In each case, the factors must be counterbalanced in order to delineate which factors seem to weigh more heavily in favour of certain countries. Although these factors are ranked by experts, they are not exhaustive and exclusive.[34] Other

[31] The IBA, *Report of The Task Force on Extraterritorial Jurisdiction*, p. 197.

[32] S. W. Brenner, 'Cybercrime Jurisdiction', *Crime Law Social Change* (2006) 46, pp. 197–204.

[33] S. W. Brenner, 'Cyber-Crime and Jurisdiction, information technology and law series,' pp. 345–349.

[34] Princeton project in universal, the Princeton principle on universal jurisdiction (2001), principle 8, p. 53. Available at: http://www.princton.edu/lapa/univ_jur.pdf (18 Aug 2013).

relevant factors should be considered in particular instances for positive jurisdictional conflicts.

Since most cyber crime such as cyberterrorism is conceptually analogous to similar traditional offences in the real world, it seems that cyber crime can be dealt with by amending the traditional penal law; there is no need to adopt penal laws which specifically target various kinds of cyber crime. However, given the physical distinction between the conduct that constitutes a cyber crime and the distinct methods necessary to constitute a break in to a computer system, it seems more logical to enact specific laws targeting cyberterrorism. Therefore, due to the inadequacy of the traditional jurisdiction principles, three theories have been developed and applied to cover jurisdiction disputes over the internet and these include: the country where uploading occurs, the country where downloading occurs and the country in which its citizens are targeted through the website.

A reasonable way to address conflict between jurisdictions is to create uniform rules which can be utilised at the international level to coordinate among states in the fight against cyberterrorism. That is, by directing or urging states to either coordinate their efforts or adopt modes of mutual recognition in cases where more than one state has an interest, instead of asking them to decide on their own to exercise jurisdiction.[35] This seems the best way to avoid jurisdictional conflict. The Convention on Cybercrime is the most important instrument in the fight against cyber crime. Although it contains specific rules about jurisdiction, it is based on the principle of territoriality. Article 22 of the Convention states: "Each Party shall adopt such legislative and other measures as may be necessary to establish jurisdiction over any offence … when the offence is committed in its territory". As has been seen, this principle is based on the traditional principle. According to the Convention principles, the place in which the criminal has committed a crime is important and this is similar to the traditional principle.[36]

---

[35] The IBA, *Report of the Task Force on Extraterritorial Jurisdiction*, p. 198.

[36] N. Foggetti, 'Transnational cyber-crime differences between national laws and development of European legislation', *Masaryk University Journal of Law and Technology* (2008) 2, p. 35.

5.5 CONCLUSION

Currently, we are at the stage where each legal system responds to the conflict of jurisdictions in cyber space with a different method, which is shown by comparative studies conducted by various scholars. Some scholars respond to this problem by borrowing personal jurisdiction and extending it to cyber space as well, while others believe that such traditional rules of private international law are inadequate to address cyber space, which does not contain any geographic indications. Therefore, it seems that the best feasible solution is providing a treaty (or convention) to regulate particular transactions to uniform international standards. This agreement can impose compulsory jurisdiction on state parties over cyberterrorism offences. This treaty can also offer a special law of a nation to be exercised on certain transactions. In a situation where no consensus can be reached, an international regulatory body can provide model law, and then the state may use it as a guide to enact its own municipal legislation.[37] However, establishing a multilateral treaty with a new harmonised and unique jurisdiction is not an easy job to perform, but it seems the most appropriate one. Furthermore, due to the global and virtual nature of cyber space, it would be best if the proposed response to this issue is determined on a multilateral basis. The best method for the prosecution of cyberterrorism under universal jurisdiction is to create a multilateral criminal law convention pertaining to cyberterrorism jurisdiction to create an affirmative obligation on member states to prosecute and extradite, because the '*aut dedere aut judicare*' principle must be established through treaty and applicable among state parties of a multilateral convention. As a matter of fact, the identity of the attacker must be determined prior to applying any of the varying kinds of jurisdiction — ranging from territorial jurisdiction to universal jurisdiction — in order to provide effective deterrence.

[37] A. M. Sachdeva, 'International jurisdiction in cyber space: A comparative perspective', *Computer and Telecommunication Law Review Oxford* (2007) 13, p. 245.

CHAPTER VI

CONCLUSION AND RECOMMENDATIONS

6.1 INTRODUCTION

A unique characteristic of cyber attacks is their ability to be launched from anywhere in the world, at any time. During the cyber attacks on Estonia in 2007, most of the compromised and attacking computers were located in the US. Cyber attacks can be set to launch under predetermined conditions or on a certain date in the future. Even if discovered, attack tools can also be difficult to remove from a computer network completely, even by forensic experts. With the latest cyber attack technology, it would appear impossible to be certain that all adversary attack options have been eliminated. The objective of this chapter is to conclude the whole issues in the research. It also presents a discussion of the findings of the research. Future research possibilities and what could have been done differently are also discussed. Finally, recommendations are put forward.

6.2 CONCLUDING ANALYSIS

Cyberterrorism is one of the greatest threats of the new century that has targeted most countries in the world. It poses so many problems to countries by producing physical and virtual damages to their national infrastructure. Countries are now combating the threat of cyberterrorism. They have provided statutes, cyber security, and joined international organisations in response to this problem.

Prosecuting cyber criminals is not possible unless investigators have the necessary legal tools to do so. The first substantial tool is a clear definition for "cyberterrorism" which covers all forms of cyberterrorism, in order to prosecute cyber attackers and to enable the use of procedural rules for evidence gathering and investigation. Due to the transnational nature of cyberterrorism, lawmakers should take such measures in their penal and procedural laws, at adequate levels, that will not allow an offender to take advantage of gaps in existing laws. Therefore, a unifying definition of terrorism and cyberterrorism is a strict requirement, in order to distinguish a cyberterrorism case from other types of cyber crime and to prosecute the attackers.

One of the main issues in responding to a cyber attack is to distinguish the origin of the attack to discover whether the attacker is a private citizen or acting on behalf of a government. This distinction is vital because it determines which body of law should control the proper response. Another major issue is the target of the attack, in order to determine available responses. For instance, the US has divided cyber attack targets into vital and non-vital targets and based on this categorisation, determines the severity of a response. The President's Commission on Critical Infrastructure Protection divides the vital targets of the US into five critical infrastructures: information and communication, physical distribution, energy, banking and finance, and vital human services. Any attack against any of these five infrastructures is considered an attack against a vital target. Any attack on anything besides these five is considered an attack on a non-vital target.

Another step in dealing with cyberterrorism incidents is the identification of the attacker. This plays an important role in ascertaining the nature of an attack, which leads to the formulation of a response to the attack. This task is not complete without the investigation and prosecution process. Countries have different regulations to investigate and prosecute cyber criminals. In cyberterrorism, there are obvious difficulties in tracking, arresting, and prosecuting cyber criminals.

Commission of such crimes presents new enforcement challenges, requiring different approaches in investigation and prosecution. The existing continuous threat of global cyberterrorism requires a response especially in ways to counter it through enforcement measures. As a new form of criminal activity that is extremely complex, cyberterrorism has to be countered through new forms of investigation and prosecution that differ from methods used in checking traditional crimes.

One issue to be faced is that the national laws as provided for by the Council of Europe Convention on Cybercrime are not harmonised. This makes it difficult to track, identify and prosecute cyber attacks and cyber criminals at the global level especially since there is no standard international convention that obliges all nations to address the new technology on which cyberterrorism is based. There are, however, international organisations that, as cross-border entities, have made efforts to combat the threat of cyberterrorism by providing a standard definition of the crime and harmonising the laws of their member states. However, they face some difficulties in curbing the threat. The first difficulty is that they do not possess sufficient authority and powers to impose sanctions on recalcitrant countries. Another problem is the lack of membership which limits the ability of international organisations to address the threat comprehensively.

The different legislations of different countries also pose a problem and some countries do not even have specific laws to deal with cyberterrorism. In the event of a cyber attack, they have to resort to traditional laws which are often inadequate to counter or handle the problem.

In most countries, no law is specifically devoted to combating cyberterrorism. These countries address the issue of cyberterrorism using current legislations on conventional terrorism and cyber crime. In using cyber crime legislation and non-internet terrorism legislation to address terrorist use of the internet, several problems arise. Another issue that was often overlooked by lawmakers was how such legislation would apply to new threats such as cyberterrorism.

The nature of the internet gives the ability to the user to disguise its identity, leading to inherent difficulties in determining the states that fail to prevent an attack that originated within their borders. Therefore, applying a proper law to cyberterrorism cases when each country is eager to implement their own law creates a confusing situation. As cyberterrorism is cross-border in nature where national borders and sovereignty do not matter, jurisdiction issues come to the fore, and scholars propose different views about the proper jurisdiction in handling cyberterrorism cases.

Following the identification of attacker, the problem of bringing the criminals to trial must be faced. Criminals are brought to trail via mutual legal assistance treaties (MLATs) and bilateral legal assistance treaties. The whole process of investigation clarifies that the existence of MLATs and bilateral and multilateral treaties among countries is vital for the capability of countries to prosecute cyber attacks. The absence of an extradition treaty between two countries renders impossible the prosecution of the attacking country. The problem is that some countries are not members of such international treaties, while in other cases the application of these treaties is so complicated and time consuming that it often renders them ineffective or difficult to apply.

Unlike cyber criminals that can operate across jurisdictional boundaries, law enforcement cannot cross jurisdictional boundaries. Law enforcement in the US and other countries are confined within their territorial boundaries for investigation and prosecution. For instance, a person who committed a crime within the US can flee the country to evade prosecution, or if he/she targeted US infrastructure from outside the US. The US has to rely on other countries' laws or existing extradition treaties. If the US does not have an extradition or legal assistance arrangement with the country where a fugitive has found haven, a barrier for investigation and prosecution is created. Extradition treaties can be complicated for a number of reasons. For instance, an illegal action in one country may be legal in another country. Such a scenario may lead to one country's reluctance to work with another or to turn over a suspect for prosecution. Another obstacle to investigations is the disparity that exists between different countries' cyber crime laws.

The main issues in addressing cyberterrorism as highlighted above are the obstacles that prevent governments from conducting an effective prosecution. Countries have been overtaken by the threat of cyberterrorism and have not been able to come up with appropriate measures to counter it such as having a specific definition of such a crime, and the fact that there is no international and unified convention for cyberterrorism which they can adopt or incorporate into their domestic legislations. To counter this they have had to use their existing laws despite the associated problems such as bringing criminals to trial on the strength of the existing laws. The advent of new technologies has created new opportunities for cyber criminals creating new situations that demand specific measures for investigation and prosecution especially due to the cross-border nature of these crimes.

6.2.1 Issues with Cyberterrorism Definitions

Cyberterrorism is a crime that is viewed as threat that is extremely dangerous. From the legal aspect, the new methods of deploying cyberterrorism attacks and the threats posed by cyberterrorism still remain unclear. This section addresses the issue of a definition of cyberterrorism as currently there is no standardised international definition that has been accepted by all nations. This leads to all the attendant problems such as those involving investigation and prosecution by affected countries. Although scholars and practitioners have come up with various definitions of cyberterrorism, there is still no consensus acceptable to all on what constitutes cyberterrorism that can cover all forms of cyber terrorist attacks.

In defining the notion of cyberterrorism we can divide the measures that compose the definition of cyberterrorism into two categories: elements originating from attack (effect) and elements originating from attacker (intent). Two groups of definitions have been established as well: the intent-based definition and the effects-based definition. Elements originating from an attacker consist of intent and motivation. Elements originating from an attack consist of effects of the attack and the target of the attack. Standard evaluations of attack allow having a clear definition in determining cyberterrorism cases. In fact, cyberterrorism incidents require substantial elements of crime and it is these elements of crime that make cyberterrorism different from other cyber crimes.

The intention of the perpetrator or the *mens rea* in deploying a cyber attack is an essential element of a crime. The *mens rea* has a central role in criminal liability in a cyberterrorism crime. The political and social motivation of a cyber attack must be sufficiently harmful and frightening to be classified as cyberterrorism. Another feature for a crime to be characterised as cyberterrorism is the target of the cyberterrorism attack, which must be aimed at national critical infrastructure. Only the definition that focuses on intention covers the full range of attacks, such as damaging economic data and releasing poison gas.

Furthermore, the attack must cause violence or severe economic and social harm. It must be the immediate cause of the effect. The effect elements must cause a sufficient amount of fear and anxiety in a civilian populace through damage to critical infrastructure. If the result of the cyber attack is equivalent to the fear and anxiety caused by traditional terrorist actions, then it will be labelled an act of cyberterrorism. Evaluating the effects of an attack enables law enforcement bodies to distinguish between real cyber terrorists and terrorists who only use the internet for other purposes that cannot be labelled as cyberterrorism.

Some definitions, as considered in Chapters II and III, fail to make this distinction clear because they do not emphasise how severe the attack must be before it is defined as cyberterrorism. If they do not focus on this point, then the cyberterrorism definition becomes so broad as to be inclusive of misconduct not generally understood as terrorism. Simply, a direct link must exist between the act of deploying a cyber attack and the result of disrupting and damaging the critical infrastructure. Potentially, all critical infrastructures are vulnerable to cyber terrorist attack. Therefore, the intent and effects of the act (individual cyber action) determine and differentiate between terrorist or criminal acts or acts of war that have occurred.[1]

Cyber terrorists' intention is to intimidate and force others through acts and threats of violence, while in cyber warfare

[1] N. Solce, 'The battlefield of cyber space: The inevitable new military branch — the cyber force', *Albany Law Journal of Science and Technology* (2008) 18, p. 293.

the combatants are trying to achieve military objectives. Generally, warfare tends not to target civilians, while terrorism is usually intended to directly or indirectly demoralise a civilian population. The primary goal of terrorism usually is to intimidate civilians by destroying property and injuring or killing civilians. A perfect example of real world terrorism is the 11 September 2001 attack on the World Trade Centre. In contrast, cyber criminals carry out cyber crimes for financial or psychological benefits, and to steal money or information, gain personal fame and attention, be intellectually challenged, and/or experience illicit pleasure.

The definition of cyberterrorism exists in the statutes of the three countries considered in this book i.e. the US, the UK, and Malaysia. They each provide a broad definition to cover all types of cyberterrorism attacks to safeguard their national security. However, in some cases they may elevate simple traditional criminal offences as constituting terrorism and cyberterrorism. One of the main differences between the cyberterrorism definition in the US and the other two jurisdictions[2] is that the US does not only define cyberterrorism in its US Code; most government agencies have also developed their own definitions of cyberterrorism, which differ from each other. Each of them has altered an element of the definition in order to fit the definition with their own goals.

Therefore, the cyberterrorism definition has become so divergent in the US as to make it a difficult task for the government to devise a strategy to combat cyberterrorism. There should be a consistent view on the basic elements of a cyberterrorism definition and a common definition must be established. There are however no inconsistencies in the definitions of cyberterrorism in the legislations of the UK and Malaysia whose laws treat terrorism and cyberterrorism as a subdivision of terrorism and do not define cyberterrorism separately.

Although cyberterrorism is a fairly new issue, a precise and certain definition of cyberterrorism will help the investigator to defend, deter, and prevent it in appropriate ways. This is because, if an accepted definition does not exist, police cannot differentiate

[2] United Kingdom and Malaysia.

cyberterrorism from other related crimes and therefore it cannot be prosecuted. The terrorism definition in the UK displays remarkable similarity to the Malaysian definition. It is the goal of terrorists to use computer technology to advance their primary goals of demoralising civilians and destabilising governments. According to the cyberterrorism and terrorism definition though (the UK, the US, and Malaysia), it is obvious that cyberterrorism and cyber-based terrorist organisations are regarded in the same light as real world terrorism. At the same time, national statutes and amendments to anti-terrorism and cyberterrorism definition laws have increased in order to upgrade the ability to confront the borderless and transnational nature of cyberterrorism as a new form of terrorism.

In conclusion, there are definitions that focus solely on one of these elements, while ignoring other elements. A combination of intent-based and effects-based definitions run the risk of making the category of cyberterrorism without difficulty and cover all the elements. It is essential for a proper definition of cyberterrorism to exist that covers all types of cyberterrorism and which has all these elements together. It will avoid the trap of not being broad enough to include other terms in its definition and not being narrow enough to exclude some cyberterrorism cases and acts.

6.2.2 The Effective Role of International Organisations in Curbing Cyber Terrorist Activities

International organisations have made efforts to combat the problem of cyberterrorism and related issues through their activities and strategies. The issue is that individual countries that are members of such international organisations have to adopt the measures they have agreed to and this raises the question of national sovereignty. Combating cyberterrorism can be achieved but it comes at the expense of interference by international organisations.

International cooperation counters cyberterrorism, leading to an effective and preventative ability to prosecute online crime and cyberterrorism offences that cross international borders. If countries want to be protected from cyber attacks and cyberterrorism, they have to cooperate, coordinate, and harmonise in international legal efforts.

To do so, such countries' coordination and harmonisation efforts are divided into two main areas. First is the legislative process combating cyberterrorism, which requires countries to legislate according to elements that are drawn from the international harmonisation effort. The second step is global investigation and prosecution of cyberterrorism and the adoption of procedural laws to prosecute criminal offences. This step should apply to the collection of evidence in electronic form of criminal offences enumerated in a list of conduct to criminalise and which has been agreed to by all countries in order to reach to a certain amount of agreement in cyber space. The various measures illustrated in this chapter indicate the need for laws to be harmonised to prevent transnational criminals from exploiting jurisdictional and legal loopholes among countries, providing fewer opportunities for them.

Although applying international law seems to be the best way to confront cyberterrorism, the difficult job of attributing old terms and norms to new technologies such as cyber space further complicates the situation. Another problem that hinders these organisations from countering cyberterrorism comprehensively is that they just offer recommendations, but do not provide sanctions for their members based on the structure of international treaties. The only internationally binding instrument is the resolution adopted by the UN General Assembly. This resolution can apply to all UN members, thereby offering an effective implementation tool. This approach is not as difficult as treaty drafting that may take long periods of time to be adopted.

International organisations are more effective in addressing cyberterrorism compared with regional organisations. The existing international conventions that work on the harmonisation of national substantive and procedure laws are applicable to the prosecution of cyberterrorism. For the moment, these international organisations seem to use the best method for preventing and prosecuting cyberterrorism cases. The Convention on Cybercrime is the most significant convention in this field that can be applied to cyberterrorism cases. The procedural law and international cooperation provision in this convention are defined broadly, going beyond cyber crime. The broad acceptance of countries is important for international instruments to fight against cyberterrorism. However, the main problem of existing conventions is the lack of signatories, ratification, and implementation. For now, two international conventions (the Convention on Cybercrime and the Convention on the Prevention of Terrorism) are the most important international instruments for combating cyberterrorism. The essential role of the Convention on Cybercrime is for the establishment of substantive laws and criminal procedures as well as laws for international cooperation.

If the UN established a specific convention on terrorists using the internet for cyberterrorism, fighting cyberterrorism would be easier. Legal measures serve a critical function in the response to cyberterrorism. Thus, having a common definition for cyberterrorism with unique characteristics will ease the establishment of cooperation among countries. Countries must achieve real cooperation by getting involved in international communities. International communities must provide conditions to promote a general consensus among countries.

6.2.3 Application of Legal Provisions in the Case of Cyberterrorism

As cyberterrorism is a new offence created by new technology, most countries do not have suitable laws to combat it and often encounter many problems when exposed to a cyberterrorism attack. Cyberterrorism mostly occurs through unauthorised access that takes advantage of the different computer laws of different countries. It is difficult to establish the required elements of unauthorised access to prove the offence and proving these requirements and their application is a rather complex process. Furthermore, the issue of convicting cyber criminals through existing provisions of affected countries is another problem, and since they often do not have specific statutes on cyberterrorism they have to utilise their existing laws to meet the threat.

A. Legal responses according to terrorism statutes

The main issue addressed in this section is two identical notions of cyberterrorism and ancillary cyber activities that should not be confused. Although both are considered as cyber activities, we cannot place them in the same category. It is a problem that cyberterrorism and terrorist use of the internet are different issues that are often confused as being the same. Differentiating between these two notions is not easy although, in combatting this menace, it is important to have a precise definition of the two activities.

Hacking tools and information obtained by terrorists from the internet are used to launch cyber attacks. Ancillary cyber terrorist activity supports terrorist objectives or the terrorists' use of the net. These acts are not characterised as cyberterrorism. The internet has introduced a platform through which hackers can inflict damage without engaging in violence. Ancillary cyber activities play vital role in deploying cyberterrorism activities. They support a cyber terrorist to do their job in an effective manner. Ancillary cyber terrorist activities, as one of the issues, are criminalised in all three jurisdiction of this book (the US, the UK, and Malaysia). All cases and terrorism related statutes in these three jurisdictions attempt to make these activities unlawful and provide harsh punishment in order to deter and prevent terrorist groups and individuals from engaging in them. In fact, cyber space is the best option for a terrorist group to deploy their attack due to the ease, low cost, speed and anonymity of the internet, as well as lack of international conventions to respond to it. Therefore, they prefer to do their job and deploy their attack through these activities.

B. Legal responses according to computer crime statutes

Most sophisticated cyber attacks have exploited the principle of information security. The goal of information security is to provide confidentiality, integrity, and availability. Most cyber attacks that have been launched exploit one or more of these attributes as they have attempted to steal information, or enable nations to spy on each other, or enable attacks on confidentiality (confidentiality attacks). These attacks can involve manipulation of data to degrade the data inside the information systems, or are designed to degrade the adversary by sabotaging the operation of critical information systems (integrity attacks) and bring information systems offline in order to prevent access to information and destroy critical physical or virtual processes.[3] All cyber attacks are deployed through various types of unauthorised access. They reach the information, data or critical infrastructure through viruses, worms, malicious code, Trojan horses, distributed denial of service (DoS) attacks, spam, spoofing and phishing, email bombing, and spyware.[4] The first viable requirement for cyber attacks and cyber exploitation (where the effect is non-destructive and refers to intelligence gathering) is access to a system or network, and finding the vulnerabilities in the accessed systems.[5] Cyber terrorists usually implement two general types of attacks: defacement of websites and DoS attacks, to launch the attacks which they make through unauthorised access. Therefore it may be important to consider the issues of computer crimes in these three jurisdictions; since a cyber terrorist attack occurs through unauthorised access.

Bearing all the above in mind, the appearance of new types of cyber crime typically invokes the passing of new criminal codes to deter and prevent the spread of cyberterrorism as a type of cyber crime. In efforts to protect their national security, the US, the UK, and Malaysia have declared unauthorised access to main computers as a criminal act through their respective Computer Fraud and Abuse Act 1984, Computer Misuse Act 1990, and Computer Crime Act 1997.

Although most countries utilise criminal law in order to respond to cyber attacks, they are not well-equipped to deal with large-scale cyber terrorist attacks. Even if the countries have joined the international convention and have implemented its provisions in their national laws, the threat of cyber attacks against critical infrastructure may remain beyond the scope of their criminal law.[6]

---

[3] D. Alperovitch, Towards Establishment of Cyberspace Deterrence Strategy, *3rd International Conference on Cyber Conflict*, Tallinn, Estonia, 2011, p. 90.

[4] H. Cheng Chu *et al.*, 'Next generation of terrorism: Ubiquitous cyber terrorism with the accumulation of all intangible fears', *Journal of Universal Computer Science* (2009) 15(12), p. 377.

[5] P. Denning *et al.*, 'The profession of it discussing cyber-attack', *Communication of the ACM* (2010) 35(9), p. 30.

[6] E. Tikk, 'Comprehensive Legal Approach to Cyber Security', Doctor Iuris Dissertation, University of Tartu, Estonia, 2011, p. 77.

6.2.4 Enforcement

Terrorism as a traditional crime committed on computer networks poses a new problem for the investigation, prosecution, and preventive of crime. The primary goal of investigation is to uncover and discover the truth. The issues of enforcement are divided into two sections in this book: investigation and prosecution.

A. Investigation

The investigation process and issue is based on digital evidence. The process of investigation is the same as for conventional crime. The dramatic increase in cyberterrorism attacks requires prosecutors to understand how to obtain electronic evidence stored in computers. New types of crimes require new laws and digital forensic investigation to enable the digital evidence to be presented in a court of law. The digital evidence is presented in the court of law and if the procedures were conducted based on the step-by-step procedure in cyber and electronic laws, it will be admissible in the court of law.

The fact is that cyberterrorism investigation is surrounded by complexity and this complexity hinders the investigators because the internet provides the opportunity for cyber criminals to react quickly. Cyber criminals utilise a distributed approach to conceal their activities across the globe and several jurisdictions. Law enforcement needs to react and keep pace with changes occurring through the different methods of cyber attacks. A key feature in conducting appropriate investigations and law enforcement is to manage information and knowledge among all parties involved. Like physical investigation, cyberterrorism investigation is information-rich and it is vital to gain a detailed insight of the illicit data attacks rapidly.

Finding the illicit intent of the cyber criminal in cyberterrorism cases is a big problem and is essential for investigators to find the illicit intent of the cyber criminal. This illicit intent forms a major part in investigating cyberterrorism, because one of the key distinctive differences between cyberterrorism and other cyber crimes is the illicit intent of the cyber criminal which must be political, religious, or similar.

Obtaining evidence and performing the investigation process in interconnected networks is almost impossible, making investigation and prosecution when dealing with networks a difficult process as well. Data and information can be deleted and altered as quickly as they are created. Perpetrators attempt to conceal their actual identities. Hence, finding the identity of the cyber attack's perpetrators is the most difficult part of the investigation process and can only be solved through cooperation among states. The first step to ascertain a perpetrator's identity is to examine the computer of the cyberterrorism victim to trace the computer that the data was sent from. Through that computer's router the network's server that launched the cyber attack can be determined. However, gathering this evidence in another country may cause problems and cause a conflict between national sovereignty and international collaboration.[7]

Cyber criminals cannot be prosecuted by law enforcement unless countries enact statutes that criminalise offenses related to terrorist activities in cyber space. Cyberterrorism footprints are fragile and transient; therefore, traditional law enforcement and investigation are not adequate. The investigator is required to obtain the assistance of the authorities of the country where the crime originated, or the countries through which the activity transited. This process is also difficult and the investigator must determine which devices are available for search and seizure of evidence which is located in other countries. Obtaining such evidence through formal mutual assistance and letter of rogatory is the next issue.

Multilateral mutual assistance treaties (MLATs) are required to obtain permission from the relevant countries' authorities for prosecution and investigation processes, or letters rogatory in the absence of a treaty or executive agreement. The procedure of MLATs is faster than the old process of letters rogatory. In the absence of an effective MLAT, the investigator will have to resort to the letter rogatory process. It is the duty of the investigator to prepare the letter rogatory and submit to the authority of the country from which they are requesting the assistance. However, obtaining assistance through a

---

[7] A. Reyes *et al.*, *Cyber-crime Investigations Bridging The Gaps between Security Professionals, Law Enforcement, and Prosecutors*, Elsevier Publication, Amsterdam, Netherlands, 2007.

letter rogatory may take a year or more; because it is transmitted through diplomatic channels which make it a time-consuming method. Investigating and prosecuting cyberterrorism and cyber crime offences requires cooperation between law enforcement officials in various countries.

B. Prosecution

As soon as the damage is established, a variety of practical and legal issues may emerge from establishing the identity of the perpetrator to obtaining evidence, and the last stage is pursuing conviction. It appears all these three jurisdictions provide extraordinary low harm requirements for acts of terrorism directed against electronic systems and other infrastructure.[8] Most countries do not distinguish between the different techniques used to launch cyber attacks, so long as the attack satisfies the more general harm and fault requirement of a terrorist act.

In common law countries, prosecutors must decide what cases they take on, while in other legal systems, prosecutors are required to prosecute when sufficient evidence is available. These decisions are reached by the prosecution policies in some modern jurisdictions based upon the seriousness of the offence and the sufficiency of the evidence, but in others, the law dictates it. The search warrant is applied in two ways in different countries. In some of them, the law enforcement officers must obtain judicial authority before searching and seizing, while in others this authority is vested in prosecutors or police. Countries that vest the search warrant in the police or prosecutor give a large amount of authority to its police, which may be abused by them. Prosecutors in common law jurisdictions must present the evidence in court; in other jurisdictions they just assist the judge in fact finding.[9] Put simply, in common law countries like the UK the prosecutors decide what cases they take on. While, according to the US' Fourth Amendment of the Constitution, in order to

---

[8] K. Hardy, 'WWWMDs: Cyber-Attacks against infrastructure in domestic Anti-Terror Law', *Computer Law and Security Review* (2011) 27, p. 153.

[9] S. Brenner *et al.*, 'Transnational evidence gathering and local prosecution of cybercrime', *John Marshall Journal of Computer and Information Law* (2002) 20, p. 347.

obtain a criminal warrant, a federal agent must prove that under the circumstances known to him or her there is a reasonable belief that a person has committed, is committing, or is about to commit a crime.

The difficulty in obtaining a search warrant is that, the requesting party or investigator must provide sufficient cause that a crime has been committed and that the electronic device located at the place was involved in committing the crime. Then the investigator must obtain a warrant from the court for search, seizure, or arrest of suspect. However, the search must be within the scope of the warrant. The digital evidence resulting from the search warrant enables the prosecution to convict criminals.[10]

Through the search warrant, an investigator can obtain personal information of suspects in a case that may involve cyberterrorism. As time is very important in police investigations, early detection and intervention is of paramount importance in counter-terrorism.[11] The problem is that in computer-related offences such as cyberterrorism, data is fragile and sensitive, and the data are at the risk of removal. This is particularly the case where the evidence is apparently scant due to encryption but where the potential results could be enormous and for cases involving national security.[12] Search warrants are applied for in two ways: by obtaining judicial authority before searching and seizing or vesting this authority in prosecutors or police. The law can be extended without sufficient reason because of the seriousness of the offence.

Among the concerns about these laws is that they may be abused by public prosecutors, for example by detaining persons without them being charged as this infringes on the privacy of the individual. The results that have been obtained from the prosecutor's action are not considered by an independent communications commission which

[10] A. Khan *et al.*, Digital forensics and crime investigation: Legal issues in Prosecution at National Level, *IEEE Computer Society*, p. 244.

[11] Terrorism Bill 2005, Clauses 23 and 24.

[12] B. B. Chatterjee, 'New but not improved: A critical examination of revisions to the regulation of investigatory powers act 2000 encryption provisions', *International Journal of Law and Information Technology* (2011) 19(3), p. 271.

has the duty to review the results and they do not have certain procedures which have to be followed by the prosecutor.[13]

C. Search warrant

The next issue after prosecution is the issue of extradition. The existence of mutual legal assistance and bilateral and multilateral treaties among countries is vital for the capability of countries to prosecute cyber attacks. Without these treaties, the whole investigation process remains ineffective. The absence of an extradition treaty between two countries renders the attacked country incapable of prosecuting the perpetrators. As an international crime, cyberterrorism provides opportunities for criminals to flee from justice by crossing borders. Judicial cooperation between nations is an urgent requirement to tackle this development. Developed countries must reform their extradition laws to contribute to that process.[14] An extradition treaty between two countries plays a significant role; particularly, in prosecuting cyberterrorism offences. This is because the right of demanding extradition from a perpetrator's country exists only in the case that is created by treaty. The main problem of the most extradition treaty is exemption of political offences from their content. Therefore, in case of computer crimes such as the one that happened between us and china, even if the extradition treaty existed, they cannot use it as they exempted political offences from their content.

Unlike cyber criminals that can operate across jurisdictional boundaries, law enforcement officials' jurisdictional boundaries are limited. Law enforcement, investigation and prosecution in the US and other countries are confined to territorial boundaries. Any attempt, for instance, to prosecute a criminal fleeing the US or targeting the US interest from outside her borders, will have to rely on the laws of the other country or on an extradition treaty existing between the two countries. If no extradition treaty or other legal assistance arrangement exists between the US and the other country, barriers for

---

[13] Available at: http://ejp.icj.org/img/cpc_terrorism_amendment.pdf (22 Feb 2011).

[14] L. J. Lioyd, *Information Technology Law*, 5th Edn, Oxford University Press, New York, 2008, pp. 272–273.

investigation and prosecution are created. Even if an extradition treaty exists, it can be complicated for a number of reasons. For instance, an illegal action in one country may be legal in another country, and may lead to a reluctance to work with another or to turn over a suspect for prosecution. Another obstacle to investigations is the disparity that exists between the cyber crime laws of different nations.

6.2.5 Rational Jurisdiction for Cyberterrorism

The main issue regarding jurisdiction in the international space of the internet is the dichotomy which exists between three components of jurisdiction in cyber space that are personal jurisdiction, territorial jurisdiction, and universal jurisdiction. Although many steps have been taken to combat cyberterrorism, from legal to technical steps, these attempts have not been sufficient to prevent cyberterrorism. It appears that greater international cooperation is needed.

The inherent and ubiquitous characteristic of the internet gives rise to the complex issue of jurisdictions. Following the identification of a cyber criminal act, the question of which international body or state will have the authority to take legal action to prosecute arises. The issue of jurisdiction seems the most problematic in the fight against cyberterrorism. The fact that cyber attack can come from anywhere around the globe makes it difficult to bring offences to the court. Cyber terrorists cross national and international borders and attack via hacking tools. Therefore, the issue of jurisdiction is important.

Various views have been presented about the issue of proper jurisdiction, and scholars differ on the type of jurisdiction that can encompass all types of cyberterrorism. In fact, individual countries prefer to apply their own legislation to offences and most of them rely on the principle of extraterritorial jurisdiction. However, this may conflict with the jurisdiction of other countries. Local regulations are not sufficient in the case of cyberterrorism which is a transnational crime and are often not adequate to check such criminal activities. Alternatively, if the principle of extraterritorial jurisdiction is applied it may violate the national sovereignty of states.

6.3 RECOMMENDATIONS

This part will put forward appropriate and suitable recommendations for the problems mentioned earlier in the concluding analysis.

6.3.1 Issues with Definition

There are several suggestions to be adopted to remedy the loopholes in the definition of cyberterrorism.

(a) States must work together closely to offer a common definition for cyberterrorism. This comprehensive definition must include the essential elements of cyberterrorism to be prosecutable by all nations. For now, every state provides different definitions about cyberterrorism with different perspectives. If states collaborate and keep certain conventions such as the Convention on Cybercrime as a basis of their cyber legislation, they can expand and harmonise their legislation according to this convention. A unified definition for cyberterrorism has to maintain imminent elements that could form a cyberterrorism attack. The main elements of this crime include the political motivation of the perpetrator. Scholars who believe in an intent-based definition of cyberterrorism focus on this aspect. While, scholars who believe in an effect-based definition of cyberterrorism focus on that aspect. However, a comprehensive definition that includes both aspects is necessary. The most complete definition combines the two elements; because, it can cover all kinds of cyberterrorism without overlapping with other related cyber crime. The crime must be accompanied by the use of violence, the use of force, or the threat of its use. This factor is accepted by 80% of scholars. The bad faith of the perpetrator is substantial. The *mens rea* of the perpetrator is an essential part of the crime that must be followed by the *actus rea* or the criminal act. The criminal act must be perpetrated using a computer and telecommunication capabilities. This attack must be against critical national infrastructure or against civilian and human life. It must be disruptive enough to generate fear comparable to a conventional act of terrorism. Such attack must result in violence, destruction of property, or civilian casualties. The property can be tangible such as critical national infrastructure. A good example of this scenario is the "Stuxnet" cyber attack that wreaked havoc in SCADA systems. Intangible properties are mostly affected by DoS attacks by changing data and information.

(b) All these elements must be covered to have a comprehensive and preventive definition. This definition should be broad and includes everything from basic hacking and DoS attacks to concerted efforts to unleash weapons of mass destruction or disruption. Such definition is also "limited in application regarding the actor or actors and the intent behind the attack".

(c) Cyberterrorism is a politically motivated crime and can cover a wide range of cyber attacks, from high-profile strikes against critical infrastructure to millions of pinprick attacks that can weaken the state over a long period of time. As for the threat of politically motivated attacks that threaten national security, they must be treated as a national security issue in order to get the full support of policy makers. However, as described previously, this would require governments to come to an agreement, the same way they did over nuclear programmes, and commit to disassociating themselves from the development or financing of these types of cyber attacks. For instance, Estonia faced these crimes by enacting its emergency management act which can be used to prevent an economic crisis by providing available responses and measures in emergency situations. Following the 2007 attacks, major changes have been enacted in Estonian legislation. Cybercrime-related provisions in Estonia's Penal Code were reviewed to harmonise them with the Council of Europe's Convention on Cybercrime which involved stiffer and higher maximum penalties.

(d) Terrorism and cyberterrorism definition in the acts of three related countries indicate that all of them have extraordinarily low-harm requirements for an act of terrorism directed against an electronic system or other infrastructure. The UK and Malaysia provide the maximum penalty of life imprisonment for all politically motivated acts of intimidation (or "influence" in the UK) that interfere with electronic systems and other infrastructure, regardless of an offender's intention to cause a greater level of harm, injury, or death. They provide broad and vague definition for terrorism; while, the US is very careful to restrict the penalty of death to the most serious acts of terrorism, and calibrate the penalties for attacks causing lesser levels of harm. Other countries can take the same approach as the US.

6.3.2 Challenges Faced by International Organisations Relating to Cyberterrorism

(a) Cyberterrorism is a new phenomenon and international norms have yet to be established. One of the main steps for advancing defences against it is to broaden international law enforcement cooperation, especially via the Council of Europe Convention on Cybercrime. However, foreign law agencies may infringe upon national sovereignty. There must be an international treaty to prohibit the development of cyber weapons, such as the treaty on chemical weapons. The articles in such a treaty might, for example, ban supply chain attacks, the disruption of non-combatant networks, and increase international management of the internet. However, a criticism to the second approach is that it does not help in cyber attack attribution.[15]

(b) As the instances indicate, a good way to combat cyberterrorism is by adopting a resolution of the UN General Assembly that is legally binding and vested by the UN Charter. This resolution would apply to all UN members, making it a strong and effective implementation tool. This happened in the case of Resolution 1373 which borrows various obligations from existing counter-terrorism conventions and applies them to all UN member states without the need for them to sign those conventions. Drafting a treaty is obviously not an easy job and may take many years to be adopted. Although not a complete international counter-terrorism instrument, creating a Security Council Resolution can be an effective tool. For example, a Security Council resolution could be adopted which covers all the elements of cyberterrorism and it should comprehensively cover the issues. This would be the most

---

[15] K. Grees, 'The Challenges of Cyber-attack Deterrence', Cooperative Cyber Defence Centre of Excellence, Naval Criminal Investigative Service, Tallinn, Estonia.

effective because such a resolution would be under the auspices of the UN of which most nations of the world are members.

(c) A serious problem of the Convention on Cybercrime lies in the lack of signing, ratification, and implementation of the parties that hinder the convention from being properly used. The number of member parties is very important, and countries must coordinate on national rules.

(d) The main problems of existing conventions (i.e. Convention on Cybercrime and Convention on the Prevention of Terrorism) are the lack of signatories, ratification, and implementation. Given that these two conventions are the most important instruments for fighting cyberterrorism, serious efforts should be made to promote the process of signing, ratifying, and implementing the convention. All additional effort pursued internationally according to these two conventions should not distract from the fact that these Conventions should be signed, ratified and implemented. Hence, amending and updating the IT aspect of the Convention on Cybercrime is preferable. Such amendments should be taken under the additional protocol. In doing so, the Convention on Cybercrime must be recognised as a basic convention, which is a normal process in the fast-paced technical environment. The Convention on Cybercrime must be updated to cover new technical advances such as new forensic investigative techniques. The convention should also exclude the political exception clause for serious offences such as cases of data and system interference, offences dealing with high risks such as cyberterrorism, and cooperation instruments for cyberterrorism cases. Legislators must adopt new provisions prohibiting serious attacks on IT-based infrastructure. Consequently, countries must evaluate their domestic statutes to make sure that they provide appropriate sanctions for offences involving terrorist attacks against essential infrastructure. However, such appropriate sanctions are already required by the Convention on Cybercrime. The result can be achieved by enacting such legislation on the national level that provides serious sentence for the offences of data interference and infrastructure damage.

(e) A protocol must also be added to the Convention on the Prevention of Terrorism to cover all illegal terrorist content, particularly content related to terrorist acts committed in cyber space.

Currently, such offences are not fully covered by other instruments of international organisations. When the threat of these offences exists, a proper response is needed.

(f) The Convention on Cybercrime was drafted in 1997, thus making its technical methods and forensic investigation tools inadequate to address suspected terrorism cases at present. The necessity of an additional protocol should be explored to update these procedural tools. The new protocol should consider the new risks posed by terrorism, new technical and forensic investigations, and development. The protocol should focus more on preventative measures to act against "the dissemination of illegal content" and its prosecution in cyber space. Adding specific control systems on the internet that will not prohibit the free exchange of information without reason can be useful. Member states need to enact domestic statutes to provide proper sanctions against illegal access and data and information interception to fulfil the convention's aims completely. The Convention on Cybercrime also requires decisive sanctions for the benefit of the member states, but authorises them to delineate such sanctions in their legislation within the convention guidelines. The large-scale attacks in Estonia and Georgia taught the world that identifying the origins of an attack will require the cooperation of other states. This objective cannot be achieved unless states are obliged through international law or duty to cooperate with other states in the investigation process.

(g) The UN should provide a specific convention to cover all issues in a systematic way. If the Convention on Cybercrime's amendments are enacted, these can be good examples for a higher level of cooperation such as a UN convention.

(h) Given the rapid changes throughout the internet, illegal terrorist content, such as those related to cyberterrorism are disseminated frequently. To promote deterrents against terrorist content and cyber terrorist attacks more effort should be made with particular attention to both illegal terrorist content in cyber space and the general way of committing cyber terrorist attacks. Rapidly changing technology affects the law: this law of cyberterrorism must also be developed rapidly along with technology. For that reason the law should amended accordingly.

(i) The importance of international cooperation was highlighted following the massive cyber attacks on Estonia and Georgia and involving Stuxnet, etc. It became much more obvious that one country can do little by itself, in responding to a cyber attack. Thus, there must be active participation in the work of organisations dealing with cyber security. This requires keeping national developments and legal frameworks up to date to serve as a foundation for new initiatives, further collaboration and regional or global forums. In addition, nations should support and encourage the ratification of instruments like the Council of Europe Convention on Cybercrime that aim to harmonise cyber crime regulations globally. But political cooperation alone is still not sufficient. Other international initiatives should be implemented for the effective prosecution and investigation of cyber crime offences such as national multilateral, bilateral and information sharing agreements, cooperation of law enforcement agencies, joint investigation teams, international exercises, and formal and informal networks.

(j) The landmark cases that have been illustrated in the earlier chapters emphasise the need for worldwide cooperation and countermeasures against cyber crime, involving major stakeholders in the public and private sectors. Countries must begin drafting cyber security strategies in the same way that the 2007 attacks in Estonia triggered the drafting of a cyber security strategy in that country. Nations must carry out a practical and wide-ranging risk assessment of their current cyber infrastructure. Furthermore, close cooperation between relevant law enforcement and other agencies is necessary, as often it is only the context and information that they can glean together which will determine if a cyber attack was initiated with criminal, espionage, terrorism, or military purposes.

6.3.3 Problems in the Application of Law to Cyberterrorism Cases

(a) Countries must evaluate existing domestic statutes that address cyber crime and anti-terrorism. They have to make sure that their legislation provides appropriate sanctions for cases involving cyber terrorist activity. These elements are already required in Article 13 of the Convention on Cybercrime. This convention requires member states to adopt “effective, proportionate, and dissuasive criminal or non-criminal sanctions”. Member states should attempt to reach this level.

(b) It is wise if other countries do the same things as European Union (EU) did. The EU enacted laws addressing issues of terrorist propaganda, recruitment, and training on the web and laws to directly confront the threat of terrorist use of websites and which can be applied to online terrorist activity such as recruitment and training. The new laws of the EU can be taken as a sign that other countries are dealing with this problem as a real issue. Therefore, it may be a wise course for other countries, particularly the US, which has encountered so many issues of cyber terrorist attacks, to do the same thing as the EU did, by introducing the new legislation to address public provocation, recruitment, and training for cyberterrorism. Because, even the Patriot Act 2001 which is latest act in this field in the US, does not say anything about online terrorist activity and fails to address the problem presented by terrorist websites.

(c) An analysis of the Estonian legal system that governs their information-based society emphasises that a secure information society is one that is thoroughly supported by rules derived from all legal disciplines. In the case of Estonia, their broad approach is revealed by their legal framework which brings together the areas of private and public law, and which includes criminal law, crisis management regulation, and wartime law or national defence legal order. Such a broad discipline was created following the 2007 attack when they amended their penal code to address the existing regulatory limitations in the application of the Code of Criminal Procedure. The Emergency Act 2009 was adopted to review the framework of national emergency preparedness and emergency management, including available responses to cyber threats. The Act can prevent an economic crisis by providing available responses and measures in emergency situations.

(d) Each State must have a complete national law to investigate and prosecute cyberterrorism as a part of cyber crime. As far as the conventional law does not cover cyber crimes, the new law must be broad enough to cover the latest cyber crimes, such as cyberterrorism. Since, most countries suffer from lacking the necessary statutes, they do not have cyber crime or anti-terrorist law, or even if they have, they have not covered cyberterrorism attacks because the laws were enacted during a time when there was no cyberterrorism cases. They also did not enact such laws with the possibility of interpreting them in a way that can cover cyberterrorism as well. The new statutes must be enacted in a way that can respond to the fast pace of technology change and the new crimes which will be created by new technology. This law also must include cross-border jurisdiction as well as extradition agreements or MLATs in order to include new types of cyber crimes, such as cyberterrorism.

(e) The domestic criminal law and other national acts of every nation should contain provisions relating to attacks against critical information infrastructure, politically motivated cyber attacks, and cyberterrorism. These should be augmented by related provisions for investigation and prosecution. Finally, the Council of Europe Convention on Cybercrime should be widely and inclusively implemented by all nations, especially when considering the cross-border nature of cyber crimes.

(f) Different legislative approaches in dealing with cyberterrorism and different legislative approaches to the issue of freedom of expression and human rights become stumbling blocks for nations working together as a cohesive unit to counter cyber attacks. Thus, countries must collaborate, debate, and learn in order to share their information and strategies in deterring cyberterrorism and thereby contributing to a more cohesive international approach to cyber security which is a basic need of all countries.

6.3.4 Issues in Enforcement of Cyberterrorism

This section will provide appropriate recommendations for the issues of enforcement which were mentioned earlier in the concluding analysis.

(a) An additional protocol should be made in the Convention on Cybercrime for the prosecution, investigation, and prevention of the dissemination of illegal content on the internet. It must include new national substantive laws, national procedural laws, and international cooperation laws. An effective prevention for the spread of illegal content in the field of substantive laws is the establishment of a harmonised criminal law with respect to illegal content, as well as harmonised rules on the responsibility of internet service providers (ISPs). Simultaneously, the necessary provision must be added for procedural laws and laws of international cooperation with special regulations on technical blocking and control mechanisms on the internet.

(b) Countries must enact adequate national criminal substantive law provisions that enable them to cover various terrorist acts. Providing special procedures for investigating criminal activity on the internet is essential.

(c) Effective investigation and prosecution at an international level requires the establishment of precise procedural regulations for collecting, preserving, and presenting evidence in electronic form via international agreements such as the Convention on Cybercrime. Mandatory data retention for ISPs for a period is required for law enforcement investigation. Member states must adopt measures for law enforcement bodies to order or obtain expedited preservation of stored computer data and expedited preservation and partial disclosure of traffic data.

(d) Cyber crimes mimic traditional criminal exploitation in general, but are executed with unprecedented ease, speed, and impact. An appropriate response to cyber crimes must be underlined by new technological disciplines that will address this new kind of crime. The new laws must be synchronised with innovative technology and must be broad enough to cover cyberterrorism which will emanate from the advent and development of new technology. The task of identifying cyber terrorists and bringing them to justice imposes impressive challenges to law enforcement agencies. This task requires a huge degree of timely cooperation. States and international organisations must cooperate with each other to create evidence preservation, to search and seize computer systems, and to develop protocols for effective investigations and prosecution.

(e) In the UK and Malaysia, the power of search warrants is granted to the police. However, it seems that it may be abused by the police officer and if this power was granted to the court it would be better. The power authorising the public prosecutor to intercept communications should be given to the court. Such authorisation should be given to persons who are accountable and under a duty to justify their decisions.

(f) In order to discourage cyber criminals from choosing countries that suffer from scant punishment as their home base the legal implications must be severe and have the same quality in all countries, regardless of the global condition. Some countries do not provide severe enough punishment, while others have strong laws and penalties that aid in discouraging such criminal activities.

(g) As the advances of cyber attacks are evolutionary, law enforcement and counterintelligence bodies must struggle to keep pace with their security implications. The government must also pay serious attention to cyber deterrence. One of the possible approaches to prevent or defend a cyber attack is deterrence. It is more effective and cheaper than other options. Other methods are dedicating far more resources, time, effort, and energy to tackle cyber crime problems. Global cyber deterrence cannot be obtained without international cooperation among countries. National legislation alone is not sufficient to deter and counter cyberterrorism action. There should be international coordination against cyberterrorism in order for it to be successful. Therefore, it seems that the best feasible solution is providing a treaty (or convention) to regulate particular transactions to uniform international standards. This agreement can impose compulsory jurisdiction on state parties over cyberterrorism offences. This treaty can also offer a special law of a nation to be exercised on certain transactions. In a situation where no consensus can be reached, an international regulatory body can provide model law, and then the state may use it as a guide to enact its own municipal legislation.

(h) Governments must allocate more funds to improve and strengthen cyber security at national, regional, and local levels. They must design a comprehensive plan at the national level that outlines all its components and which is coordinated to establish the most effective system for cyber deterrence. They must collaborate to counter the broad range of threats at the international level. It will be better if cyber crime units and law enforcement cooperate at the international level, so as not to be limited by national borders. It must not be forgotten that a large number of obstacles have existed at the international level for countering cyberterrorism; since there is a large amount of mistrust between countries, this leads countries to believe that they should not share information related to their national security and internal policies.

(i) Governments should commit to procedures to disable future participation in cyberterrorism while pledging to investigate and punish responsible parties. In fact, the perpetrators do their best to remain anonymous. They often use Chinese or Russian servers to appear to show that the attack originated from there, but actually, they are not located in those countries. It is crucial that the government and law enforcement agencies cooperate actively with foreign investigators and give them the freedom to investigate in their own jurisdictions in case they do not have the capabilities to do so by themselves.

(j) According to Brian Carrier and Eugene Spafford, the Integrated Digital Investigation Process Model (IDIP) is the most suitable model of investigation for cyber crime and cyberterrorism. The model covers the process from collection of digital evidence to presentation in court. It includes both physical and digital crime investigation. The IDIP model consists of four phases. The first phase is the Readiness Phase which ensures that the human capacity and underlying infrastructure of the phase are well-equipped and sufficient in order to deal with sudden incidents. The second phase is the Deployment Phase, which detects the incident and notifies appropriate people and they confirm the incident, then obtain authorisation for legal approval to carry out a search warrant to detect the incident and carry out further investigation.

By authorising the search warrant, the Collection and Analysing Phase emerges. The third phase is the physical crime scene investigation, which is aimed at collecting and searching the documents by preserving the crime scene and capturing and recording as much information as is possible. The second and third phases, that is the investigation stage, include searching and identifying evidence on a computer, then collecting that evidence, or storing the evidence collected at the scene and transferring it to a secure environment. Prior to analysing the value of the evidence, the evidence is examined with the proper tools.

Finally, the evidence is analysed and presented to the court or legal entities for investigation. At this level, the analysis must be presented and the hypothesis reached during investigation must be proved. The other phase is the Digital Crime Scene Investigation Phase, where the aim is to collect and analyse the digital evidence that was obtained from the physical evidence. Thus, all the processes carried out for obtaining information in a physical investigation are applied to digital investigation as well.[16] In the Trace back Phase, the investigator traces and identifies the devices that were used for committing the criminal act.

The previous phases assist investigators to obtain clues to trace the primary crime scene (and the suspect). For instance, they can acquire IP addresses and locate the country and institution, and host computer. They can then obtain authorisation for further investigation. The last phase is the Dynamite Phase, which collects and analyses the information obtained from the primary crime scene (the physical and digital crime scene investigation) that will eventually lead to identifying the culprit. The pieces of information from the digital and physical scenes are then put together and this gives a hypothesis. Consequently, the final interpretation and conclusion is given to the court. Ultimately, as mentioned above, the IDIP model seems to be the best model for cyber crime investigation and has the capabilities

---

[16]V. Baryamureeba *et al.*, The enhanced digital investigation process model (2004) Institute of Computer Science, Makerere University. Available at: http://makerere.ac.ug/ics. (14 Apr 2012).

to address cyberterrorism because it requires "digital investigation to address issues of data protection, data acquisition, imaging, extraction, interrogation, analysis, and reporting".

6.3.5 Jurisdiction Issues

(a) The best means for the prosecution of cyberterrorism under universal jurisdiction is to create a multilateral criminal law convention that will oblige member states to prosecute and extradite offenders through the '*aut dedere aut judicare*' principle established through the treaty and applicable to state parties of the convention. As a matter of fact, the attacker must be identified prior to the application of any of the varying kinds of jurisdiction — ranging from territorial jurisdiction to universal jurisdiction — so as to provide effective deterrence.

INDEX

www.ingramcontent.com/pod-product-compliance
Ingram Content Group UK Ltd.
Pitfield, Milton Keynes, MK11 3LW, UK
UKHW020400150926
8624IPUK00004B/258

9 781786 342126